Vision Zero at Work

This book introduces Vision Zero, a commitment to create workplaces free from serious accidents, harm and work-related diseases while delivering excellence in Safety, Health and Wellbeing (SHW).

Based on the principle that a safe and healthy working environment is a fundamental labour right, Vision Zero integrates SHW aspects to promote understanding and deliver safer and healthier workplaces. It places people at the core of business practices and offers an innovative perspective on achieving sustainable workplace safety and health. Readers will discover the foundational values, concepts and theories that define Vision Zero, along with actionable strategies for a proactive preventive workplace culture. The book emphasizes the importance of shifting from compliance to continuous improvement and innovation which is driven by leadership commitment from above and workforce collaboration from below. Practical insights on integrating prevention measures during the early stages of planning and design ensure SHW challenges are addressed before they arise. Real-world examples and policies from international organizations further demonstrate the power of Vision Zero in action, making the concept applicable to any workplace. Readers will obtain the knowledge and tools to create safer, healthier and more resilient organizations through the Vision Zero concept.

Vision Zero at Work: Strategies for Sustainable Excellence in Safety, Health and Wellbeing is essential for line managers, leaders and professionals in workplace safety, health and wellbeing or in human resource management, as well as policymakers, consultants and organizational change agents in business and management.

Vision Zero at Work

Strategies for Sustainable Excellence in Safety, Health and Wellbeing

Gerard I.J.M. Zwetsloot

CRC Press
Taylor & Francis Group
Boca Raton London New York

CRC Press is an imprint of the
Taylor & Francis Group, an **informa** business

Designed cover image: Shutterstock

First edition published 2025
by CRC Press
2385 NW Executive Center Drive, Suite 320, Boca Raton FL 33431

and by CRC Press
4 Park Square, Milton Park, Abingdon, Oxon, OX14 4RN

CRC Press is an imprint of Taylor & Francis Group, LLC

ISBN: 978-1-041-02682-2 (hbk)
ISBN: 978-1-041-02862-8 (pbk)
ISBN: 978-1-003-62122-5 (ebk)

DOI: 10.1201/9781003621225

Typeset in Times
by Newgen Publishing UK

Contents

Foreword by Hans-Horst Konkolewsky xi
Acknowledgements xiii
About the Author xv
List of Abbreviations xvii

Chapter 1 The Global Challenge of Safety, Health and Wellbeing at Work 1

1.1 Vision Zero and the Right to a Safe and Healthy Working Environment 1
1.1.1 The Importance of Work and Employment 1
1.2 The Global Human Toll of Work 2
1.3 The Interactions between the Fundamental Labour Rights 3
1.4 The Economic Dimension 6
1.5 Trends and Developments Globally 7
1.6 Conclusions 8

Chapter 2 The Changing World of Work and SHW 11

2.1 Work and Production in a World of Change 11
2.1.1 Economic Changes 11
2.1.2 Technological Changes 12
2.1.3 Demographic Changes 12
2.1.4 Ecological Changes 13
2.1.5 Cultural Changes 13
2.1.6 Change Invokes Increasing Uncertainties 14
2.2 The Impact of Change on Socio-Technical Systems 14
2.2.1 The Technological Subsystem 15
2.2.2 The Organizational Subsystem 16
2.2.3 The Human Subsystem 17
2.2.4 The Interactions between the Three Subsystems and the Organizational Environment 17
2.2.5 Technology–People Interactions 18
2.2.6 Technology–Organization Interactions 19
2.2.7 Organization–People Interactions 19
2.2.8 Socio-Technical Systems, Continual Change and SHW 20
2.3 Safety, Health and Wellbeing in a World of Change 21
2.3.1 SHW Management as a Dynamic Process 23
2.3.2 Resilience 24
2.3.3 The Absence of Negatives 24
2.3.4 Positive Approaches to SHW 26

2.3.5 SHW as Resources for Work and Productivity 28
2.3.6 Collaborative Safety and Health 30
2.3.7 SHW as Human and Social Values 31
2.4 Dealing Adequately with Change 32
2.5 Conclusions 32

Chapter 3 Vision Zero 36

3.1 What Is Vision Zero? 36
3.2 The Three Roots of Vision Zero 38
3.3 The Four Key Challenges for Vision Zero 39
3.4 The Business Perspective 40
3.4.1 The Strategic Value of Vision Zero 41
3.4.2 SHW as Business Opportunities 42
3.4.3 Putting Vision Zero on the Business Agenda 43
3.4.4 The Resource View on SHW and Good Business 44
3.4.5 The Business Case/Benefits and Costs of SHW Improvement 47
3.5 Goal Zero 50
3.6 Innovating to Zero 53
3.6.1 The Seven Innovative Qualities of Vision Zero 54
3.7 The Commitment Strategy 55
3.8 The Zero Debate 58
3.8.1 Critical Arguments That Helped to Strengthen Vision Zero 58
3.9 Conclusions 59

Chapter 4 Vision Zero: Underlying Concepts and Theories 63

4.1 Values That Support Vision Zero 63
4.1.1 The SHW Supporting Values 64
4.2 Aligning and Sharing Values 66
4.2.1 Aligning Values 66
4.2.2 Sharing Values and Giving Them Concrete Meaning 68
4.2.3 The Broader Meaning of the SHW Supporting Values 71
4.3 Ownership, Intrinsic Motivation and Purpose 72
4.3.1 Internal Motivation and Purpose 72
4.3.2 Conducing Ownership and Intrinsic Motivation 74
4.4 Doing the Right Things Right 76
4.4.1 Doing Things Right 76
4.4.2 Doing the Right Things 76
4.5 Proactivity 79
4.6 Mental Models in Times of Change 81
4.6.1 The Roots of Our SHW Mental Models 82

4.7 Learning ... 86
4.8 Conclusions ... 90

Chapter 5 Leadership: The Art of Using Power Wisely ... 94

5.1 A CEO's Perspective on Leadership and Vision Zero ... 94
5.1.1 What Is Your Motivation for 'Zero'? ... 94
5.1.2 What Are Your Main Achievements? ... 95
5.1.3 What Made a Difference? ... 95
5.1.4 What Is Your Takeaway for Business Leaders? ... 96
5.2 Senior Leadership and the Challenge of Control ... 96
5.3 The Importance of Leadership ... 97
5.4 Personal Commitment ... 98
5.5 The Employment Relationship and Mutual Commitments ... 101
5.5.1 Two Visions on Employment Relations ... 101
5.5.2 The Psychological Contract ... 102
5.6 Promoting and Enabling Vision Zero ... 104
5.7 Challenges for Leadership ... 105
5.7.1 Receiving Adequate SHW Information ... 105
5.7.2 Balancing SHW and Productivity ... 107
5.7.3 Preventing Drift to Danger ... 108
5.7.4 Promoting and Enabling Innovation and Change ... 109
5.7.5 Dealing Positively with Critical SHW Events ... 110
5.7.6 The Communication between Managers and SHW Professionals ... 111
5.8 Limitations of Leadership ... 112
5.9 Conclusions ... 113

Chapter 6 From Traditional to Proactive Prevention and Promotion ... 117

6.1 The Traditional Hierarchy of Prevention Measures ... 117
6.1.1 The Limitations of the Hierarchy of Prevention ... 119
6.2 Primary, Secondary and Tertiary Prevention ... 120
6.2.1 Limitations of the Differentiation between Primary, Secondary and Tertiary Prevention ... 121
6.3 SHW Promotion ... 121
6.3.1 Limitations due to System Level ... 123
6.4 Summarizing the Confusion about Traditional Prevention ... 123
6.5 Proactive Prevention ... 124
6.5.1 Proactive Prevention in the Early Stages of Development and Design ... 126
6.5.2 Design Thinking ... 127
6.5.3 Precaution ... 128
6.5.4 Proactive Prevention and Work Organization ... 129
6.5.5 The 12 Principles of Worker-Friendlier Design ... 129
6.5.6 Proactive Prevention during Production and Maintenance ... 134

6.6 Digitalization and (Proactive) Prevention 135
6.7 Conclusions 136

Chapter 7 Prevention Culture 140

7.1 What Is a Prevention Culture? 140
7.2 Formal and Informal Prevention 142
7.3 A Strong Prevention Culture 145
7.4 The Importance of Communication 148
7.4.1 Social Marketing of SHW – Sending the Vision Zero Message 148
7.4.2 Two-Way Communication 149
7.4.3 The Importance of Feedback 149
7.5 Barriers to a Strong Prevention Culture 150
7.6 Stages of Development 153
7.7 Conclusions 154

Chapter 8 Strategies for Long-Term Continual Improvement 157

8.1 Optimization and Innovation 157
8.1.1 Optimization 157
8.1.2 Innovation 159
8.2 Underlying and Supporting Issues and Processes 160
8.2.1 Legal Requirements for SHW Risk Management 160
8.2.2 Standardization 161
8.2.3 Adapting to the State of the Art 163
8.2.4 The Deming Cycle: Plan, Do, Check, Act 164
8.2.5 Human Error and SHW Behaviour 166
8.2.6 Creating Synergies 167
8.2.7 Economic Incentives 167
8.3 Optimization and Innovation (II) 168
8.4 Conclusions 169

Chapter 9 Strategies for Practical Implementation of Vision Zero 171

9.1 The Seven Golden Rules 171
9.1.1 Golden Rule 1: Take Leadership – Demonstrate Commitment 173
9.1.2 Golden Rule 2: Identify Hazards – Control Risks 173
9.1.3 Golden Rule 3: Define Targets and Develop Programmes 174
9.1.4 Golden Rule 4: Ensure a Safe and Healthy System – Be Well Organized 174
9.1.5 Golden Rule 5: Ensure Safety and Health in Machines, Equipment, Materials and Workplaces 175

9.1.6 Golden Rule 6: Improve Qualifications – Develop Competence 175
9.1.7 Golden Rule 7: Invest in People – Motivate by Participation 176
9.2 Proactive Leading Indicators 177
9.2.1 The Challenge of Finding Good Indicators 177
9.2.2 Leading Indicators 178
9.2.3 The 14 Proactive Leading Indicators 179
9.2.4 The Complementarity of the Proactive Leading Indicators to ISO 45001 183
9.3 Vicious and Virtuous Processes 184
9.3.1 Vicious Processes and How to Stop Them 184
9.3.2 Initiating Virtuous Processes 184
9.4 Dealing with Weak Signals and Early Warnings 187
9.4.1 Weak Signals, Complexity and Emergence 190
9.5 Using the Practical Wisdom of the Workforce 190
9.6 Starting with Vision Zero after a Fatal or Serious SHW Event 191
9.7 The Don'ts of Vision Zero 192
9.7.1 Examples of Avoiding Doing the Wrong Things 193
9.7.2 Examples of Too Much of a Good Thing 193
9.8 Conclusions 194

Chapter 10 Broader Perspectives 197

10.1 Broader Synergies 197
10.1.1 Synergies between Vision Zero and Quality Management 197
10.1.2 Synergies between Vision Zero, Sustainability and Corporate Social Responsibility 199
10.1.3 Synergies with Other Social Goals 202
10.2 National and International Policies for Vision Zero 203
10.2.1 National and International Regulation 205
10.2.2 Labour Inspectorates and the Promotion of Vision Zero 207
10.3 Vision Zero in the Value Chain 208
10.3.1 Safety Health and Wellbeing Plus (SHW+) 208
10.3.2 Developing SHW in the Value Chain 210
10.4 Societal Impacts of Vision Zero 211
10.5 Conclusions 212

Chapter 11 Evidence for Vision Zero 215

11.1 The Challenge of Evidence 215
11.2 Limitations of Present SHW Research 217

11.3 Key Characteristics of Vision Zero That Are Relevant for Research and Evidence....218
11.4 Lines of Research Conducive to Vision Zero....219
11.5 Conclusions....221

Index....225

Foreword

It is with great honour and pleasure that I introduce this important and most timely book *Vision Zero at Work: Strategies for Sustainable Excellence in Safety, Health and Wellbeing*, authored by Professor Dr Gerard I.J.M. Zwetsloot, a pioneering figure in the development and promotion of the Vision Zero approach to workplace safety.

This comprehensive work represents not only a culmination of Professor Zwetsloot's decades of expertise in the field but also an essential contribution to the ongoing evolution of safety, health and wellbeing (SHW) practices across the globe.

The Vision Zero strategy is rooted in the unwavering belief that every worker has the right to a safe and healthy working environment and that no work-related injury or illness is acceptable. It calls for a systemic, transformative approach to safety that goes beyond compliance and risk management to a broader, cultural shift where prevention is proactive, integrated and continuously improved. This is not a theoretical concept but a practical framework for organizations aiming to eliminate harm and promote a healthy, productive workforce.

In his book, Professor Zwetsloot distils years of research, analysis and real-world experience into actionable insights that guide organizations towards achieving zero harm. The book begins with an exploration of the global challenges surrounding workplace safety and health. It examines the multifaceted economic, social and human impacts of work-related injuries and illnesses, which continue to take a significant toll on workers, industries and societies at large. It is clear that safety health and wellbeing are no longer just an operational concern but a fundamental moral and economic imperative.

The book emphasizes that Vision Zero is not merely a set of guidelines or a set of lofty ideals but a strategic business approach that brings profound, long-term benefits. The elimination of harm in the workplace contributes to improved reputation, employee satisfaction, organizational performance and innovation. In Chapter 3, the essence of Vision Zero is presented, highlighting its roots in zero defects, safety culture and legislative initiatives. The ambitious objective of 'Goal Zero' is discussed as a guiding principle – not as a target in itself, but as a continuous, aspirational journey towards excellence.

As the report progresses, it delves deeper into the core concepts underlying Vision Zero, particularly the values of proactivity, ownership and intrinsic motivation that must be embedded within the culture of any organization seeking to implement it. Chapter 4 discusses how organizations can align and share these values, creating a shared sense of responsibility among all workers. This transformation requires a shift in mental models and an ongoing process of learning, ensuring that every stakeholder has a role in driving improvement.

One of the standout features of Professor Zwetsloot's work is his emphasis on leadership. In Chapter 5, the report explores the critical role of leaders in promoting and enabling Vision Zero within organizations. Leadership goes beyond setting rules or policies; it is about inspiring a cultural change and fostering an environment where safety is prioritized and integrated into daily work. Leaders must balance competing interests, navigate challenges and engage their workforce to ensure that Vision Zero is not a top-down directive but a collective commitment.

The book also challenges traditional approaches to prevention, emphasizing the need for proactive strategies that are embedded into work processes from the outset. Chapter 6 critiques conventional methods that often focus on reactive measures, like behavioural controls and protective equipment, and advocates for a more integrated approach. Proactive prevention involves considering risks at the design stage, using innovative technologies and ensuring that feedback loops between production, maintenance and design contribute to continuous improvement.

Finally, the book offers practical tools and strategies for the implementation of Vision Zero, with concrete steps for applying proactive leading indicators, managing risk and engaging the workforce in the process. From the application of the seven Golden Rules to identifying virtuous processes, the report provides a comprehensive roadmap for organizations at any stage of their journey towards Vision Zero.

Professor Zwetsloot's vision of safety, health and wellbeing at work is nothing short of transformative. This book is not only a reflection of his expertise but also an invitation for all organizations to embark on a journey of change, with the goal of achieving zero harm and fostering a culture of care, responsibility and innovation. By embracing the Vision Zero philosophy, we can make the workplace a safer, healthier and more fulfilling environment for everyone.

It is my hope that this book serves as a catalyst for change in organizations worldwide and that it inspires leaders, safety professionals and workers to come together in the pursuit of Vision Zero because every worker deserves to return home from work as healthy and safe as when they arrived.

Thank you, Gerard, for your immense contribution, not just in renewing our understanding but in driving a profound paradigm shift in prevention policies and practices. Your dedication has inspired us all, and we look forward to continuing this journey together.

Hans-Horst Konkolewsky
President
International ORP Foundation
December 2024

Acknowledgements

It is my habit to see similarities where others tend to see differences: similarities between safety, health and wellbeing at work (SHW); similarities between the management of SHW, quality management and environmental management (the topic of my doctoral thesis); Similarities between SHW and corporate social responsibility; and so on. Only after I completed my PhD, I realized this was a personal quality that I should further explore. That generated opportunities to work with many great people from different disciplines. This was inspiring and taught me a lot. Similarities often imply relationships and opportunities for synergies and organizational learning.

Like Bob Dylan already knew, the times are always changing. As a researcher, you never know in advance whether your work will be appreciated, or not. A few times I was lucky, and there was a great interest in my work. In other periods there was hardly any broader interest in what I regarded as important findings. That experience made me modest, but it also made me aware that the present global interest in Vision Zero for SHW implied a kind of implicit invitation to write a book on Vision Zero. As many aspects of Vision Zero interact, it took me quite some time to find a useful structure for the book. Inevitably, many aspects described in the various chapters and paragraphs are closely related or interdependent. I am sure the readers will discover and explore such relationship further while reading the book. Hopefully, that will give them further clues for practical improvements.

This book builds on my professional experience developed during decades of work. It has been a privilege to be able to work with so many great people. Too many people to name them – all have contributed to my understanding of SHW opportunities and the challenges of Vision Zero. It is with gratitude that I remember some inspiring colleagues at the Dutch Safety Institute and the Netherlands Institute for the Working Environment, NIA. I will never forget the relevance of their practical knowledge and experience. As part of the team *Safe and Health Business* at the Netherlands Organization for Applied Scientific Research TNO, I worked with many inspiring colleagues. These included Johan Gort, Annick Starren, Linda Drupsteen and Johan van der Vorm (mostly for safety), Anja Dijkman and Arjella van Scheppingen (primarily on health management) and Remco Visser and Mat Jongen mainly on working with chemical substances.

I am very grateful for the support of Jacques Allegro and Frank Pot who made it possible to combine my work at TNO with that at Erasmus University Rotterdam and Nottingham University. That implied opportunities to work with wonderful and renowned colleagues and a few talented PhD students. I would like to thank Donald Huisingh and Leo Baas of Erasmus University Rotterdam for introducing me to the world of cleaner production and sustainability. I owe a lot to my Nottingham University colleagues Tom Cox, Aditya Jain and Stavroula Leka and our European partners in the PRIMA-EF project where we further developed the management of psycho-social risks and wellbeing at work.

There are also many interesting people I co-operated with on a project basis. The cooperation with Nick Ashford from MIT and Frank Nijman (who shortly thereafter joined Golder Associates in Indonesia) on inherently safer production was a great experience. I want to thank all the PEROSH colleagues who contributed to our projects on the Zero Accident Vision and its implementation, especially Markku Aaltonen and Pete Kines.

I am particularly grateful to the many organizations that participated in our research or consultancy projects. Their leaders, workers and SHW professionals shared their challenges and knowledge and gave access to their organizations or networks. These were sometimes invaluable experiences. It has been a great inspiration and pleasure to work for the Special Commission on Prevention of the International Social Security Association (ISSA), especially with Bernd Treichel, Helmut Ehnes and Sigrid Roth.

I want to thank the people who gave feedback on the draft chapters of the book: Herman de Bruine, Linda Drupsteen, Jan Michiel Meeuwsen, Frank Pot, Victor Roggeveen, Annick Starren, Arjella van Scheppingen and Johan van der Vorm. Thanks also to Matteo Marocco for his help with the figures. I am grateful and honoured that Hans-Horst Konkolewsky was willing to write the foreword.

Finally, Lia Karsten was and is the anchor point and inspiration in my private life. Jointly we managed to combine a good family life with the development of innovative scientific perspectives. Lia's concept of the 'back-seat generation', the generation that is overprotected by their parents and as a result is more vulnerable than resilient, provides food for thought about growing up, education and the importance of safety, resilience and wellbeing for people and society.

Gerard Zwetsloot
Amsterdam, 11 January 2025

About the Author

Gerard I.J.M. Zwetsloot is an independent researcher and consultant based in Amsterdam, Netherlands. He specializes in workplace safety, health and well-being, with a focus on Vision Zero and prevention culture. Previously, he served as a senior researcher at TNO and held honorary professorships at Erasmus University Rotterdam (1998–2006) and Nottingham University (2007–2019). He has a BSc in chemistry and physics, an MSc in chemistry (Utrecht University) and a PhD in social sciences (Leiden University). Gerard has led key projects for the International Social Security Association (ISSA) and its global Vision Zero strategy, including the development of the proactive leading indicators guide. He is widely published and frequently invited as a keynote speaker at international conferences. In 2019, he received the Mukaidono Safety Award from the Institute of Global Safety Promotion in Tokyo for his contributions to Vision Zero.

Abbreviations

AI	Artificial Intelligence
ALARA	As Low as Reasonably Achievable
ALARP	As Low as Reasonably Practicable
BAT	Best Available Technologies
BATNEEC	Best Available Technologies Not Exceeding Excessive Costs
CaE	Culture as Espoused
CaL	Culture as Lived
CEO	Chief Executive Officer
CMR	Carcinogenic, Mutagenic and Reprotoxic (chemicals)
CSR	Corporate Social Responsibility
EFQM	European Foundation for Quality Management
ESG	Environmental Social and Governance (criteria)
EU	European Union
EU OSHA	European Agency for Safety and Health at Work
FMEA	Failure Mode and Effect Analysis
FMRA	First Minute Risk Assessment
GDP	Gross Domestic Product
GP	General Physician
GR	Golden Rule(s)
HRO	High Reliability Organization
IALI	International Association of Labour Inspectors
IEC	International Electrotechnical Commission
ILO	International Labour Organization
ISSA	International Social Security Association
IT	Information Technology
LMRA	Last Minute Risk Assessment
LTI	Lost Time Injury
MaD	Management as Done
MaI	Management as Imagined
MBO	Management by Objectives
NGO	Non-Governmental Organization
NIOSH	National Institute for Occupational Safety and Health (United States)
OE	Operational Excellence
OECD	Organization for Economic Co-operation and Development
ORP	Occupational Risk Prevention (Foundation)
OSH	Occupational Safety and Health
PDCA	Plan, Do, Check, Act – the Deming cycle
PDSA	Plan, Do, Study, Act – the Deming cycle as promoted later by Deming
PLI	Proactive Leading Indicator
RA	Risk Assessment
RCT	Randomized Controlled Trial

REACH	Regulation on the registration, evaluation, authorization and restriction of chemicals (European Union)
SDG	Sustainable Development Goal (United Nations)
SHW	Safety, Health and Wellbeing at Work
SHW+	Safety, Health and Wellbeing and labour and human rights relevant for SHW
SMART	Specific Measurable, Achievable, Realistic and Timely
SME	Small and Medium-sized Enterprises
TQM	Total Quality Management
TWH	Total Worker Health (NIOSH)
UK	United Kingdom
UN	United Nations
UNICEF	United Nations Children's Fund (originally: United Nations International Children's Emergency Fund)
VZ	Vision Zero at work
WaD	Work as Done
WaI	Work as Imagined
WCM	World Class Manufacturing
WHO	World Health Organization
WFD	Worker Friendly Design
ZAV	Zero Accidents Vision

1 The Global Challenge of Safety, Health and Wellbeing at Work

1.1 VISION ZERO AND THE RIGHT TO A SAFE AND HEALTHY WORKING ENVIRONMENT

In 2022, after several years of discussion, the International Labour Organization (ILO) – the tripartite United Nations (UN) organization (global employers, trade unions and governments) – recognized a safe and healthy working environment as a fundamental principle and right at work.[1] That principle and right is at the heart of Vision Zero. It is the right to always return home safely and healthy after work.

The right to a safe and healthy working environment implies the moral duty for employers and workers to create and maintain a working environment that is good for safety, health and wellbeing (or mental health) at work. Safety, health and wellbeing at work (abbreviated in this book as SHW) are important to everybody who is working to earn a living. In a more indirect way, good SHW is important to every employer: it is good for productivity and ensures compliance with legislative requirements in the area of occupational safety and health. On a societal level, good SHW contributes to a flourishing economy and to the welfare of our communities. Unfortunately, this does not mean that SHW is always high on the national or business agenda.

1.1.1 The Importance of Work and Employment

Work and employment are important for people. Work generates income; it implies social contacts and opportunities to use and develop competencies and can imply opportunities for personal growth. Work can be meaningful for the individual, for the organization and for society at large. Employment is important in many ways. It is usually better to be employed than to be unemployed. To emphasize the positive potential of work, the Finnish Institute of Occupational Health uses the mantra *Wellbeing through Work*. Indeed, good work can contribute significantly to the wellbeing of people at work and to society at large. A safe and healthy working environment is a necessary condition to realize the positive potential of work. Unfortunately, the health and social-economic impacts of employment are dramatic, especially from a global perspective.

DOI: 10.1201/9781003621225-1

1.2 THE GLOBAL HUMAN TOLL OF WORK

Every day thousands of people die due to unsafe or unhealthy working conditions, and many more suffer from work-related injuries and diseases, each individual case having a great impact on the quality of life of the victims and their families and communities.

The ILO regularly estimates the global human toll of work. There is no global register for documenting work-related deaths. In several parts of the world, there is no reliable recordkeeping of SHW data, and even in developed countries the data can be incomplete. Registration is often unreliable as the notification of a work-related death may imply a penalty for the employer. The ILO figures for the global human toll are therefore the best estimates available; they tend to be conservative estimates.

In 2022, it was estimated that around 2.9 million workers die annually due to occupational accidents and diseases.[2] On an average day, more than 7,900 people die as a result of unhealthy and unsafe working conditions. Unhealthy working conditions caused 81% of the work-related deaths (i.e., around 2,350,000 deaths per year, 6,400 per day) and unsafe working conditions 19% (550,000 per year, 1,500 per day) (see Figure 1.1).

The two main factors known to cause work-related death are *long working hours* (i.e., structurally more than 55 hours per week) which causes around 745,000 deaths annually – often due to stroke – and *exposure to particulate matter* (including asbestos), *gases and fumes* resulting in around 450,000 deaths annually.[3] Only since the WHO/ILO data of 2022 has it been recognized that structurally long working

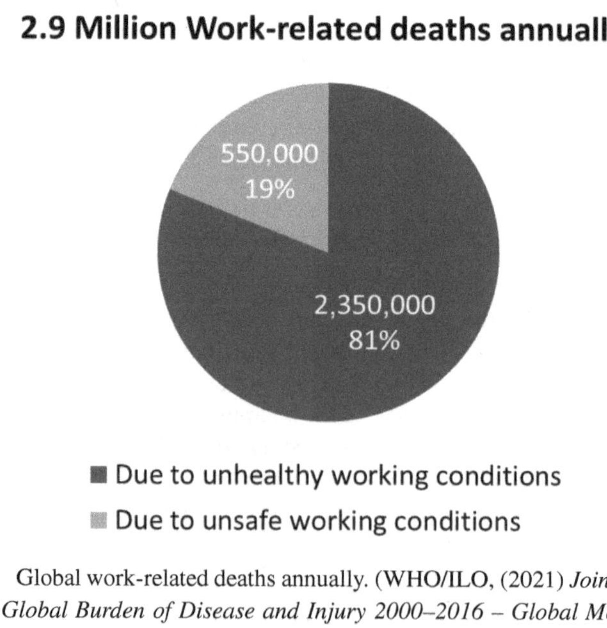

FIGURE 1.1 Global work-related deaths annually. (WHO/ILO, (2021) *Joint Estimates of the Work-related Global Burden of Disease and Injury 2000–2016 – Global Monitoring Report.* Geneva: World Health Organization and International Labour Organization. Available at: www.who.int/publications/i/item/9789240034945 (Downloaded 10 November 2022).

hours are known to be the most important cause of work-related death. This remained unknown for a long time as long working hours can result in stroke, making it difficult to distinguish cases of work-related stroke from general cases of stroke. Perhaps other important causes of work-related illnesses or deaths will be identified in the future. Part of the occupational fatalities are caused by work-related road traffic. Work-related motor vehicle crashes are estimated to contribute significantly to work-related deaths, but to avoid counting them twice (as traffic deaths and work-related deaths) these figures are in many countries not included in registers of work-related deaths. In the US, in 2022 fatal roadway incidents caused 25% of the total work-related fatal injuries.[4]

The ILO also estimates that more than 400 million people suffer annually from non-fatal occupational injuries that cause absence of work for at least four days. The WHO and ILO jointly estimated that almost 90 million disability-adjusted life years were attributable to 41 pairs of occupational risk factors and health outcomes.[5] According to the Global Risk Poll of Lloyds Register Foundation and Gallup (2019), 19% of workers worldwide reported to have been injured seriously at some point while working, which concerns around 600 million people.[6] Almost a quarter (24%) of people worldwide said they or someone they know had experienced serious harm from the work they did.[7] Nevertheless, the risks of occupational hazards are frequently underestimated, as is the severity of these risks. Of course, the frequencies of work-related accidents and diseases differ per region, country and sector. Generally, these frequencies are higher in low-income countries than in high-income countries, but there are big differences between sectors and regions within most countries.

1.3 THE INTERACTIONS BETWEEN THE FUNDAMENTAL LABOUR RIGHTS

The right to a safe and healthy working environment was recognized only in 2022 as a fundamental principle and labour right. Four other fundamental labour rights were already recognized earlier. Table 1.1 shows the five fundamental labour rights.

TABLE 1.1
The five fundamental labour rights

- the effective abolition of child labour
- the elimination of all forms of forced or compulsory labour
- the elimination of discrimination with respect to employment and occupation
- freedom of association and the effective recognition of the right to collective bargaining and
- a safe and healthy working environment.

Source: ILO, (2022) *A Safe and Healthy Working Environment Is a Fundamental Principle and Right at Work*. Geneva: International Labour Organization. Available at: www.ilo.org/publications/safe-and-healthy-working-environment-fundamental-principle-and-right-work (Downloaded 17 November 2022).

The five fundamental labour rights are closely related, which means that the absence of one of the other fundamental labour rights is also a threat to a safe and healthy working environment.

The right to a safe and healthy working environment is closely associated with the abolition of child labour and the elimination of forced labour. It remains often implicit that child labour almost by definition implies violating the right to a safe and healthy working environment and has detrimental effects on the development of the children and their SHW. UNICEF estimates that around 160 million children were involved in child labour in 2020.[8] This accounts for almost 10% of the children worldwide. Almost half of them are involved in hazardous work that directly endangers their health. In the poorest countries in the world, around 20% of the children (ages 5–17) are engaged in child labour. The occurrence of child labour, especially hazardous child labour, decreased significantly between 2000 and 2012, but it has remained more or less constant since then, with a tendency to increase since 2016.[9]

Similarly, forced or compulsory labour implies poor working conditions, often affecting the safety, health and wellbeing of working people. The estimates of forced labour amounted to 27.6 million workers globally on any given day in 2022.[10] Of these 27.6 million workers, 15.8 million are male (boys and men), and 11.8 million are female (girls and women); more than 3.3 million are children younger than 17 years. The global number of people in forced labour is increasing, *not* decreasing; in 2016, it was estimated at 24.9 million people, implying an increase of more than 10% in the period between 2016 and 2022. Forced labour is found in all parts of the world, with almost half of the people working in forced labour in Asia and the Pacific.[11] The highest proportion of forced labour is found in the Arab states: 5.3 per thousand people. About 86% of forced labour occurs in the private sector, and 14% is state-imposed. Migrants are especially vulnerable to forced labour. As opposed to child labour, the majority of forced labour is found in upper-middle and high-income countries – only 16% in low-income countries.

Non-discrimination at work is another fundamental labour right. Unfortunately, there is no recent reliable data about discrimination at work. According to the ILO, hundreds of millions of people suffer from discrimination in the world of work.[12] Discrimination in the labour market or at work is a violation of human and labour rights and clearly affects wellbeing at work significantly. As discriminated people are overrepresented in the least attractive and most dangerous and unhealthy jobs, it is also associated with poor SHW.

Freedom of association is important to balance the power inequality between employers and individual employees. Freedom of association is closely associated with the quest for the political emancipation of workers and democracy. It is a prerequisite for collective bargaining which is an essential process to achieve fair wages and fair working conditions, including SHW. In many parts of the world, there is a decline in unionized workers. This is partly due to the fact that the classic distinction between employer and employees gets blurred by platform work, the gig economy and globalization. Precarious workers can then be said to be independent entrepreneurs while they are still depending heavily on their main client who in this way can avoid the responsibilities of an employer.

Indeed, it is not difficult to see that SHW is directly associated with each of the four other fundamental labour rights and that the absence of these rights is often a cause of poor SHW. Nevertheless, the fundamental labour rights are seldom mentioned by researchers or consultants as causes of poor SHW.

The fundamental labour rights are fully in line with the Universal Declaration of Human Rights,[13] wherein two related universal work-related rights are mentioned: receiving an adequate living wage[14] and avoiding extremely long working hours (defined by the ILO as longer periods with more than 55 hours per week).[15] Table 1.2 shows the articles of the universal declaration of human rights that imply work-related rights.

The close relationship between labour and human rights on the one hand and SHW on the other hand emphasizes that SHW can be approached from an ethical or rights perspective. From that perspective, SHW is closely related to the Corporate Social

TABLE 1.2
Articles of the universal declaration of human rights and labour rights

Article	UN Declaration of Universal Human Rights – articles relevant for SHW at work	Labour rights mentioned
2	Everyone is entitled to all the rights and freedoms set forth in this Declaration, without distinction of any kind, such as race, colour, sex, language, religion, political or other opinion, national or social origin, property, birth or other status. Furthermore, no distinction shall be made on the basis of the political, jurisdictional or international status of the country or territory to which a person belongs, whether it be independent, trust, non-self-governing or under any other limitation of sovereignty.	No discrimination
4	No one shall be held in slavery or servitude; slavery and the slave trade shall be prohibited in all their forms.	No forced labour, No child labour
23	Everyone has the right to work, to free choice of employment, to just and favourable conditions of work and to protection against unemployment.	Safe and healthy working conditions
	Everyone, without any discrimination, has the right to equal pay for equal work.	Equal pay
	Everyone who works has the right to just and favourable remuneration ensuring for himself and his family an existence worthy of human dignity, and supplemented, if necessary, by other means of social protection.	Living wage
	Everyone has the right to form and to join trade unions for the protection of his interests.	Freedom of association
24	Everyone has the right to rest and leisure, including reasonable limitation of working hours and periodic holidays with pay.	Avoiding extremely long working hours

Responsibility (CSR) of organizations. From a CSR perspective, workers and their representatives form an important stakeholder group that has a material interest in, and influence on, the organization. The ethical perspective is also increasingly important to investors; they increasingly apply Environmental, Social and Governance criteria in the selection of the organization they want to invest in.[16]

At first sight, the fundamental labour rights seem not very relevant in the high-income countries. But in many high-income countries, there are significant social-economic differences and under-privileged workers. In almost all countries, there are (illegal) immigrant workers, informal work and other forms of precarious work, e.g., gig work. Exploitation, discrimination and non-compliance to the other fundamental labour rights may occur. Moreover, the fundamental labour rights are relevant from the perspective of SHW in (global) supply chains (see Section 10.3).

1.4 THE ECONOMIC DIMENSION

Accidents and ill-health imply human suffering, but they also imply costs for the employer, the workers involved and society at large. The costs for the employer include costs for medical treatment, compensation, insurance, liabilities, production losses (e.g., due to disturbances in production, sickness absence, increased turnover, material damage or time for accident investigations) and costs for maintaining production (such as overwork, recruitment and training of temporary personnel). Moreover, it may cause a loss of reputation, thereby affecting the relationships with customers, the labour market, governmental agencies and investors. Costs for

FIGURE 1.2 The symbol of SDG no. 8: Decent work and economic growth.

society at large include costs for social security, health care, disabilities and early pensioning.

At a global level, the economic losses are calculated at 5.4% of the global Gross Domestic Product. In most countries, the losses range between 3 and 7% of GDP, with 9 or 10% for a few individual countries.[17] This implies that there are good economic reasons to significantly improve SHW and that improvements in SHW can go together with economic growth. It is really a socio-economic challenge. The relationship between decent work conditions and economic growth is a tight one and is also made explicit in the UN Sustainable Development Goal (SDG) number 8: *Decent Work and Economic Growth* and it symbol, see Figure 1.2.

Vision Zero and the quest for good SHW is relevant for three of the other SDGs as well, namely SDGs no. 3: *Good Health and Well-being*, SDG no. 9: *Industry, Innovation and Infrastructure* and SDG no. 12: *Responsible Consumption and Production*. The relationships of SHW with the UN SDGs once more emphasize the link between sustainability and CSR.

1.5 TRENDS AND DEVELOPMENTS GLOBALLY

Globalization and technological developments have contributed to major economic changes. Around the turn of the century, Western countries like the USA and the EU have off-shored many traditional industries towards low-income countries. Many dangerous jobs were exported this way, associated with a shift towards more service industries and a decrease of traditional SHW risks in the higher-income countries. In this way, better SHW statistics were obtained, even when no real progress in prevention was achieved. For a number of reasons, off-shoring production is no longer very attractive. Now, stagnation in the improvement of SHW in the West can be observed, as is clear from a recent EU OSHA report that says

> Stagnation can be observed in important areas. Since 2005, the share of workers exposed to traditional safety and health risks – accidents, noise, vibrations, dust, chemical and biological agents, high or low temperatures, electrical shock and so on – remains at a stable level. Also, ergonomic risks – repetitive hand-arm movements, tiring and painful positions, lifting and carrying, and prolonged sitting – can pose major health risks, and the statistics show no significant decrease.[18]

Though we have much more knowledge about SHW than several decades ago, this has only resulted in partial improvements. New technologies have introduced new SHW risks. The shift towards more service industries and brain work as well as the intensification of work and the erosion of the standard employer–employee relation have led to a sharp increase in work-related mental health problems, which in this book are regarded as challenges to wellbeing.

In each country, there are companies that have a relatively good SHW performance. Often these are frontrunner companies that want to get rid of accidents and work-related illnesses and diseases, sometimes they belong to a multinational company. However, often the progression in SHW performances is slowing down in industries as well, whereby the frequencies of accidents are levelling off, while mental health problems are sharply in the increase. The challenge of Vision Zero, of creating excellent SHW, is therefore relevant for organizations around the globe.

The sections above show that in each region or country, many people suffer from work-related injuries or illnesses. The fundamental principle and right to a safe and healthy working environment clearly remains a pipe dream for many working people. This has significant social and economic consequences. Industries as well as governments are therefore motivated to improve SHW significantly in their organizations and jurisdictions. But the data show that the current approaches and policies have limited effect only, and a new impetus is badly needed. Since 2017 when a global strategy and campaign was launched to promote Vision Zero globally, Vision Zero has shown to spark such impetus. Vision Zero and the associated processes of continual SHW improvement and innovation are now the challenge in the area of work and SHW.

1.6 CONCLUSIONS

A healthy and safe working environment is a fundamental labour right. Nevertheless, more than 7,900 people die each day as a result of unhealthy or unsafe working conditions. Around 400 million people suffer annually from non-fatal accidents that cause absence of work for at least four days. Each individual case has a great impact on the quality of life of the victims and their families and communities. How many people are suffering seriously from work-related illnesses and diseases is unknown, but certainly the number is huge.

The five fundamental labour rights are closely related. The absence of one of the other fundamental labour rights is also a threat to SHW. Around 160 million children were involved in child labour in 2020. Child labour has remained more or less stable since 2016, with even a tendency to increase since 2016. Almost half of these children are involved in hazardous work that directly endangers their health. The global number of people in forced labour is increasing, not decreasing; migrant workers are especially vulnerable. Discrimination in the labour market or at work affects wellbeing at work significantly and is often associated with poor safety and health. Freedom of association is important for collective bargaining which is an essential process to achieve fair wages and fair working conditions, including SHW. Western countries like the USA and the EU have off-shored many traditional industries towards low-income countries. Many dangerous jobs were exported this way. In this way, better SHW statistics were obtained, even when no real progress in prevention was achieved. Unfortunately, progress in safety, health and wellbeing as well as in each of the fundamental labour rights is stagnating or worse.

Accidents and ill-health imply human suffering, but they also imply costs for the employer, the workers involved and society at large. The costs for the employer include costs for medical treatment, compensation, insurance, liabilities, production losses and so on. Costs for society at large include costs for social security, health care, disabilities and (too) early pensioning. On a global level, the human and social losses of work amount to 5.4% of the world's GDP, ranging between 3 and 7% in most countries. As a response, the UN SDG number 8 is Decent Work and Economic Growth.

Both social and economic impacts of poor health and safety are unacceptable, especially as these losses are preventable. Industries as well as governments are therefore motivated to improve SHW significantly in their organizations and jurisdictions. Proactive prevention should be much higher on the agenda of governments, institutions, organizations as well as individual employers and workers. Through the challenge of Vision Zero, active prevention becomes a core principle for institutions and organizations.

NOTES

1 ILO, (2022) *ILO Declaration on Fundamental Principles and Rights at Work (1998), as amended in 2022*. Geneva: International Labour Organization. Available via https://normlex.ilo.org/dyn/nrmlx_en/f?p=1000:62:::NO:62:P62_LIST_ENTRIE_ID:2453911:NO (Accessed 3 March 2025).

2 ILO, (2022) *Enhancing Social Dialogue towards a Culture of Safety and Health*. Geneva: International Labour Organization. Available at: www.ilo.org/resource/enhancing-social-dialogue-towards-culture-safety-and-health-presentation (Downloaded 9 October 2022).

3 ILO, (2023) *A Call for Safer and Healthier Working Environments*. Geneva: International Labour Organization. doi: 10.54394/HQBQ8592. Available at www.ilo.org/sites/default/files/wcmsp5/groups/public/@ed_protect/@protrav/@safework/documents/publication/wcms_903140.pdf (Downloaded 24 March 2025).

4 US Bureau of Labor Statistics, (2024) *Injuries, Illnesses and Fatalities*. www.bls.gov/iif/latest-numbers.htm (Accessed 22 August 2024).

5 WHO/ILO, (2021) *Joint Estimates of the Work-related Global Burden of Disease and Injury 2000–2016 – Global Monitoring Report*. Geneva: World Health Organization and International Labour Organization. Available at: www.who.int/publications/i/item/9789240034945 (Downloaded 10 November 2022).

6 LRF-Gallup, (2019). *The Lloyds Register World Risk Poll, Full Report 2019 Poll*. Lloyds Register Foundation. Available at: https://wrp.lrfoundation.org.uk/2019-world-risk-poll/ (Accessed 8 March 2023).

7 LRF-Gallup, (2019). *The Lloyds Register World Risk Poll, Full Report 2019 Poll*. Lloyds Register Foundation. Available at: https://wrp.lrfoundation.org.uk/2019-world-risk-poll/ (Accessed 8 March 2023).

8 UNICEF, (2024). *What Is Child Labour?* www.unicef.org/protection/child-labour (Accessed 20 November 2024).

9 ILO and UNICEF (2021). *Child Labour. Global Estimates 2020, Trends and the Way Forward*. New York and Geneva: United Nation's Children Fund (UNICEF) and International Labour Organization (ILO). www.ilo.org/publications/major-publications/child-labour-global-estimates-2020-trends-and-road-forward (Accessed 10 November 2022).

10 ILO/WF/IOM, (2022) *Global Estimates of Modern Slavery – Forced Labour and Forced Marriage*. Geneva: International Labour Organization, Walk Free and International Organization for Migration. Available at: www.ilo.org/global/about-the-ilo/newsroom/news/WCMS_855019/lang--en/index.htm (Accessed 14 November 2022).

11 ILO/WF/IOM, (2022) *Global Estimates of Modern Slavery – Forced Labour and Forced Marriage*. Geneva: International Labour Organization, Walk Free and International Organization for Migration. Available at: www.ilo.org/global/about-the-ilo/newsroom/news/WCMS_855019/lang--en/index.htm (Accessed 14 November 2022).

12 ILO, (2007) *Equality at Work: Tackling the Challenges*. Available at: www.ilo.org/publications/equality-work-tackling-challenges-global-report-under-follow-ilo-0 (Accessed 20 November 2024).

13 ILO, (2022) ILO Brief. *Setting Adequate Wages: The Question of Living Wages.* Wcms 857678. Geneva: International Labour Organization. Available at: www.ilo.org/resource/brief/setting-adequate-wages-question-living-wages (Accessed 20 November 2024).

14 Jain, A., Leka, S., Zwetsloot, G.I.J.M., (2018) The fundamental human right to health, safety and well-being, in Jain, A., Leka, S., Zwetsloot, G.I.J.M., (eds), *Managing Health, Safety and Wellbeing, – Ethics, Responsibility and Sustainability.* Dordrecht: Springer Science, pp. 139–174.

15 United Nations, (1948) *Universal Declaration of Human Rights.* Paris: General Assembly of the United Nations. www.un.org/en/about-us/universal-declaration-of-human-rights (Accessed 16 November 2022).

16 See especially Chapters 4 and 5 in: Jain, A., Leka, S., Zwetsloot, G.I.J.M., (2018), *Managing Health, Safety and Wellbeing – Ethics, Responsibility and Sustainability.* Dordrecht: Springer Science.

17 Takala, J., Hämäläinen, P., Saun, R., Nygard, R.H., Gagliardi, D., Neupane, S., (2024) Global-, regional- and country-level estimates of the work-related burden of diseases and accidents in 2019. *Scandinavian Journal of Work Environment and Health* 50 (2), 73–78. Doi: 10.5271/sjweh.4132.

18 EU OSHA, (2023) *Occupational Safety and Health in Europe: State and Trends 2023,* Bilbao: EU OSHA. Available at: https://osha.europa.eu/en/publications/occupational-safety-and-health-europe-state-and-trends-2023 (Accessed 15 May 2023).

2 The Changing World of Work and SHW

The global challenges described in Chapter 1 are strongly influenced by the context. There are several major global developments (economic, technological, political, etc.) that influence work and production in all parts of the world, directly or indirectly. Some are positive for SHW, others have mainly negative impacts on SHW. One factor is, however, increasingly regarded as a constant factor: *change*. Continuing changes in production and work imply that SHW at work is changing as well. Consequently, traditional approaches to SHW are no longer always adequate. From time to time even successful practices should be reconsidered: are they still the best option given the new developments in technology and society?

Though we have much more knowledge about SHW than several decades ago, this has only resulted in limited improvements, mostly in occupational safety and traditional health issues. As was shown in Chapter 1, the toll of work is still huge. New technologies imply opportunities for SHW but have introduced also new SHW risks; the shift towards more service industries and brain work implies a decrease in safety hazards and physical burden, but jointly with the intensification of work it has led to a sharp increase in work-related mental health problems – here regarded as challenges to wellbeing.

Change triggered the development of a range of new views and approaches for SHW, each with certain strengths and limitations. A concise overview helps to make use of the complementarities and potential synergies of these approaches.

2.1 WORK AND PRODUCTION IN A WORLD OF CHANGE

Organizations interact with, and are influenced by, changes in the context, e.g., developments in economy, technology, demographics, ecology or culture. As we will see, these changes trigger changes in organizations and in SHW.

2.1.1 Economic Changes

The increase in global trade has normalized that products are made from materials and components that originate from various parts of the world. Global value chains depend on a global division of work, i.e., labour in various parts of the world.

DOI: 10.1201/9781003621225-2

While in the past most industries worked with local or regional people, major industries are now attracting talented people from all over the world. Activities that are regarded as non-essential are outsourced to whoever can do the work better or cheaper via contracting and subcontracting, often implying offshoring to developing countries or making use of migrant workers in precarious jobs.

Due to the economic pressure and global competition, work has intensified. This implies higher pressure to be productive, greater work pressure and growing psychosocial risks. As a result, there is a sharp increase in mental health issues. It also implies fewer chances on the labour market for people who are less productive (e.g., due to a disability, an illness, involvement in informal care or their age).

2.1.2 Technological Changes

The ongoing rapid developments in information technology continue to have disruptive impacts on production processes, work and SHW in all sectors. Industry 4.0 and Society 5.0 are concepts that imply radical changes through the integration of IT in almost all aspects of work and production through tools, machines and clothing that mutually communicate, produce masses of (real-time) data and are empowered with artificial intelligence. In the 'internet of things', tools, machines, workplaces and clothing communicate with each other and with a central platform. The increasingly powerful IT increases the ability to control processes and workers, even from a distance. This offers the capability to centrally control much more workplaces and people than before, and consequently it decreases the possibilities for bottom-up initiatives. It has triggered a renaissance of scientific management and (over)specialization[1] and generated platform work and the gig economy. See Section 6.6 for the potential of digitalization for prevention.

Thanks to IT, citizens have access to much more information than ever before. Platforms like Google and Facebook have detailed information about almost every person in the world (associated with privacy issues) and have successfully developed new global business models and new types of jobs. However, it has also brought a decline in the status and prevalence of many occupations that were once based on knowledge or access to information.

2.1.3 Demographic Changes

The main global demographic changes are an increase in migration, diversity, life expectancy (ageing) and growing inequalities. Migration is increasing for two main reasons: (1) skilled workers see better opportunities in the global labour market elsewhere and (2) people who try to survive under hard conditions in their home country, e.g., due to high levels of unemployment, conflicts, poverty, failing government or ecological disaster migrate in the hope to find a better place to live and work. Both developments trigger growing cultural diversity in organizations and societies. Due to different values, religions and languages, this makes communication and cooperation within organizations more challenging. Thanks to better living conditions and health care, life expectancy is growing in many parts of the world. It has led to ageing populations in societies, ageing workforces in many organizations and the tendency to increase the age of pensioning, with some exceptions for burdensome occupations. Consequently, in many organizations there is a mix of generations at work, which implies differences in education, experience and associated cultural values. Due to the ageing workforce, chronic public and welfare diseases have entered the workplace, implying that public health issues impact workplaces and the productivity of companies. Managing health can no longer focus only on the prevention of the more traditional physical occupational illnesses and diseases but also needs to address mental health and consider the impact of welfare diseases.

2.1.4 Ecological Changes

Ecological change and the need for a transition towards (environmental) sustainability have triggered a rise in ecological and social concerns in society. The UN SDGs and Vision Zero are both examples of such concerns. The societal acceptance of environmental risks and risks for health and safety is gradually decreasing.

The dramatic potential of climate change triggers sustainable innovation, including new business models. The increasing use of electrical energy instead of fossil fuels implies benefits such as lighter tools and less exhaust gasses in the workplace. But climate change implies more extreme weather (heat, storm, heavy rain and severe drought), and endangers workers in great parts of the world. According to a recent ILO report,[2] the impact of climate change on the health and safety of workers is dramatic, as a result of excessive heat, more ultraviolet radiation, extreme weather events, increasing workplace air pollution and vector-borne diseases and greater use of agrochemicals. Climate change implies harder living and working conditions in many parts of the world, affecting SHW and triggering poverty and migration.

Limited stocks of scarce materials as well as climate change trigger transitions towards the use of renewables and the development of more circular economies. Treating waste as a source of valuable materials implies not only the development of new business models (e.g., leasing instead of selling products) but also a rise in dirty and dangerous work in waste treatment and recycling activities.

2.1.5 Cultural Changes

The complexity and abstract nature of many problems and risks lead to a growing dependency of society on the knowledge of experts. As there are often economic or political consequences associated with their decisions, Ulrich Beck[3] predicted already in 1992 a growing mistrust towards experts, scientists and science – a phenomenon that we can now clearly observe. Knowledge and information are omnipresent. Algorithms play a huge role in the information that is presented to people. It has become difficult to distinguish fakes from facts and complot theories from science. This development is enhanced by experts/researchers who do not make their underlying assumptions and uncertainties explicit. As a consequence, their work is being criticized as being no more than opinions or social constructs, and the status of scientists is decreasing further. This process is strengthened by companies and institutions that deliberately undermine science, by hiring scientists to purposely neglect existing or upcoming problems (e.g., about long-term health or environmental impacts) to defend their vested economic interests.

The meaning of work is also changing. In the past, making money was of prime importance for almost every worker, but that is gradually changing. Work is still important to generate an income for the overwhelming majority of workers, but for an increasing number of people, money is no longer the primary motivator for work. Depending on the type and organization of work, work implies opportunities for social contact, creativity, learning and personal growth. Partly independent of income, work also co-creates status in society. Young people are increasingly eager to become successful workers and make a great career as an expression of their self-realization. But organizations often have a performance-oriented culture and offer increasingly temporary jobs; that combination causes a lot of stress and uncertainty, especially for young workers. Organizations expect their workers to be continuously high-performing. At the same time, employment is more uncertain due to the proliferation of precarious work. For many (young) workers, work life is a kind of rat race. This is a threat to SHW, especially for mental health.

2.1.6 Change Invokes Increasing Uncertainties

Rapid changes and increasing complexity are associated with growing uncertainties. Organizations are uncertain about economic developments, their (future) profitability, their survival and (potential) new legal requirements. Today's best practices may no longer be the best practice tomorrow. The global 'war for talent' implies an increase in uncertainties about being able to select and retain talented and qualified personnel.

People have to deal with more uncertainties as well: uncertainties about their position in the labour market, career perspectives and the tenability of their qualifications; about their functioning in continuously changing organizations and multicultural teams; about social security in case of setback; and about how to be able to combine a meaningful job with intensification of work, a pleasant private life and being able to stay healthy and work till the age of pensioning and beyond.

2.2 THE IMPACT OF CHANGE ON SOCIO-TECHNICAL SYSTEMS

Organizations are socio-technical systems. Usually, three different subsystems are thereby distinguished: the functioning of technological, social/organizational and human subsystems (see Figure 2.1).

The understanding of social-technical systems and the interactions of the subsystems are hampered by the limitations of the various disciplinary silos and their associated mental models. Engineers tend to have blind spots in their understanding of social and human processes and how they interact with technology; social scientists tend to have blind spots in their understanding of technological processes; and so on. Human resource professionals and physicians tend to think in terms of individuals and their behaviour and often have difficulties in understanding a 'systems approach' or the interactions with technology. Lawyers are often more interested in finding

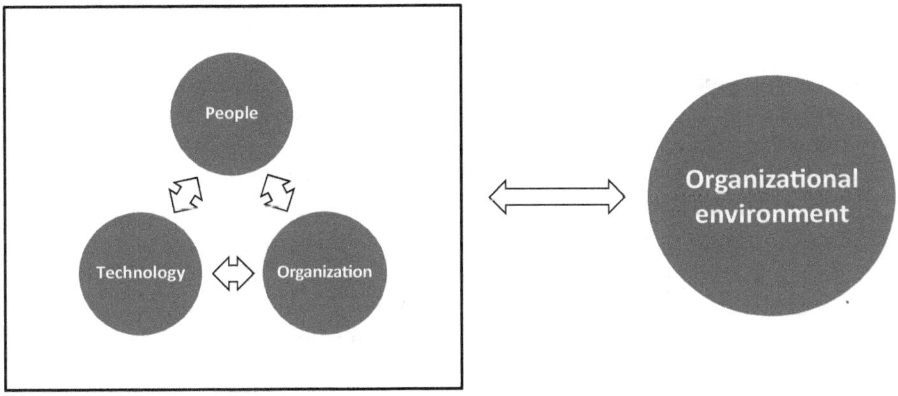

FIGURE 2.1 The interactions between socio-technical systems and the organizational environment.

somebody guilty or preventing liabilities than in understanding the socio-technical system that generates or facilitates the risks and behaviours at stake.

Technology, organization and people are always associated with certain hazards and risks, as well as with solutions and opportunities for improvement. Prevention measures can be effective by influencing or controlling hazards and risks associated with the three main subsystems: technology, organization and people and their interactions. We take a closer look at the three subsystems before we take a look at the impact of the changing organizational environment and the interactions between the three subsystems.

2.2.1 The Technological Subsystem

Around us, we see drastic changes in information technology which impact the economy, organizations and our way of working, living and communication. Digitalization is the key word for industry 4.0, thereby making use of, e.g., automation, artificial intelligence (AI), big data, digital twins, machine learning, robotics, sensors, virtualization or telework. Sensors and wearables provide new types of data and strengthen the power of big data. IT-driven autonomous systems are being developed, e.g., autonomous driving of cars and trains. E-learning and virtualization play an increasing role in education. Digital twins play a role in the design of complex systems.

The rapid proliferation of IT raises a new series of ethical questions. What implicit values and assumptions codetermine the functioning of algorithms? Who is liable or accountable in case the use of artificial intelligence leads to undesired results? Inequalities in digital competences also lead to growing socio-economic inequalities.

Though the word 'technology' is nowadays often associated primarily with IT, there are many other (production) technologies which also imply opportunities, threats and new uncertainties for SHW. This is true, e.g., for the rapidly developing areas on nanotechnology, new materials, biotechnology, and new energy carriers.

TECHNOLOGICAL COMPLEXITIES

Technology and its interactions with people and the organization is an inseparable part of any organization. Quite a few social scientists are of the opinion that technology is relatively simple and predictable, in contrast with the social aspects.[4] They thereby have a blind eye to technological dynamics and complexities. The simplified view of these social scientists on technological issues is probably partly the consequence of the fact that engineers do their very best to design technological systems that are reliable and function in a predictable way. Tools are nowadays increasingly designed human-friendly, while the instructions for use are usually written for dummies (also because producers want to avoid liabilities for an inadequate or unintended use). However, that does not mean that technological complexities do not exist or are irrelevant.

It is not necessary to dive into quantum theory to find technological complexities. A single pendulum is very predictable (and is therefore often used as a mechanism in clocks). But any double pendulum is still a very simple purely mechanical system, but it acts chaotic and fully unpredictable. Many chemical reactions are influenced by minor components (contaminants) that may serve unexpectedly as catalysts, which influence the speed of chemical reactions, often resulting in several less-desirable by-products. In the production of so-called 'fine chemicals' (e.g., in the pharmaceutical industry), it is not uncommon that 95% of the products are by-products. Software programmes are notorious for their unforeseen vulnerabilities; the development of software programmes seems to mirror a continuous cat-and-mouse game between software developers and potential hackers. The safety impact of software in safety-critical systems is often not depending on the reliability of the software as such, but on complex interactions with the wider system design.[5] A biological example of complexity is the process of mutation whereby the genetic properties of an organism are spontaneously altered (e.g., the various variations that became dominant during the Covid-19 pandemic).

2.2.2 The Organizational Subsystem

Organization can be designed according to a variety of paradigms, mental models or 'images.'[6] Each image leads to specific strengths and weaknesses. Morgan's 'images' are a kind of ideal types, but in practice the machine-like organization is still dominant in many of today's organizations. A machine-like organization is designed by the top and supported by experts. It is a rational design wherein causes and effects are assumed to be clear. It typically has a strong hierarchical approach and a high degree of specialization (Taylorism). Planning is important to guarantee control and predict results, as is the standardization of processes. The mechanical organization is conducive to SMART (Specific, Measurable, Achievable, Realistic and Timely) planning of activities. Discipline of workers is important; they need to work according to clear, unambiguous rules prescribed by management. Though the machine-like organization makes it difficult to adapt to new circumstances, many companies like the qualities of it. It is therefore not surprising that many organizations also manage SHW in a rather mechanistic way. SHW management mirrors the organizations they are part of.

A high degree of specialization implies a strong division of labour. The underlying assumption is that specialization leads to higher efficacy and efficiency of work processes and so to higher productivity and profit. But a high level of specialization also implies more interfaces: when people have to hand over the result of their work to the next colleague, or when people from different specializations have to work together. In practice this is an important source of problems; it generates problems of communication and cooperation; it leads to human and organizational errors and inefficiencies which often remain unnoticed and passed on in the organization; it triggers defective products, reduces productivity and profit and generates SHW risks. It is also a main internal source of complexity and the emergence of unexpected phenomena and risks.

The opposite of Taylorism is the organizational design according to the so-called 'socio-technical school', whereby division of labour is kept minimal and teamwork is facilitated. Jobs are then comprising a diversity of tasks, including planning and regulation of work.[7] That type of socio-technical design of organizations leads to more opportunities for learning on the job and better SHW. More in general design of the work organization has a strong influence on the functioning and development of work, on the functioning of companies and institutions as a whole and on SHW.

2.2.3 The Human Subsystem

It is a bit odd to speak about humans in terms of systems. Every human being is unique. No two human beings are the same. People differ in genetics, cultural backgrounds, intelligence, health status, social skills, experiences, gender, age and so on. Not only individuals but also teams are unique; in the process industry, it is well-known that every shift has its own characteristics. Nevertheless, the functioning of people collectively can be regarded as a kind of human subsystem. In organizations, individuals are operating in a social-economic context; they have colleagues, bosses and people they influence and are influenced by. They produce goods and/or services with economic value.

Some of the relevant individual characteristics or human qualities relevant for work and SHW are attitudes, beliefs, commitment, competence, creativity, critical thinking, emotional intelligence, flexibility, habits, intelligence, knowledge, motivation, persistence, responsibility, skills, talents and values. There are also characteristics and qualities of groups of people, e.g., in teams or organizations: communication, cooperation, culture, leadership, mutual learning, participation, social support and values like trust. Interpersonal interactions in teams or organizations are almost by definition ambiguous and therefore difficult to plan, describe or analyse objectively.

2.2.4 The Interactions between the Three Subsystems and the Organizational Environment

The functioning of social-technical systems cannot be understood solely by understanding each of the three subsystems. The interactions between the subsystems are often more important than their functioning as such. Some interactions in the socio-technical system have a direct impact (in safety jargon: the subsystems are tightly coupled), and other interactions have a less straightforward and often more long-term impact (loosely coupled).

The interactions of people with technology and with organization, as well as the interactions between technology and organization, are key for the performance of any socio-technical system. Unfortunately, all these interactions are to a certain degree ambiguous, multi-interpretable and therefore difficult to manage.

2.2.5 Technology–People Interactions

In socio-technical systems, the interactions of technology with organizational arrangements and human behaviour are essential. The human–technology interaction is always rather complex. Even a simple aspect, e.g., the right way to fine-tune an office workplace (computer, keyboard, screen), may be a challenge, especially in case of desk-sharing. The simple tool of 'five times why' easily clarifies that problems in this area are not only a matter of behaviour of the workers. They are also influenced by the technological properties of the tools and the workplace and by the organizational factors like culture, leadership and policies.[8]

FIVE TIMES WHY

A simple and well-known tool to combine several perspectives and to identify relevant relationships associated with problems within socio-technical systems is the five times because tool. Starting with a very concrete (usually undesirable) event and asking five times why helps to understand the multi-causal nature of many problems in socio-technical systems. What seems to be a 'human error' at first sight may then be linked to a problematic design of a technological tool, or an organizational arrangement. These problematic designs may be triggered by a lack of management attention. Lack of management attention may be caused by the mental models of the senior managers (perhaps based on past successes).

Global ICT Platforms have become very powerful and created new types of work in the gig economy. ICT platforms tend to strengthen power differences. Platform workers may enjoy some flexibility but have very little say in the organization or performance of their work. The relationship between platform and worker is seldom transparent, while power distances are enlarged. Artificial intelligence has the potential to make many jobs redundant and increase the distance between those at the top and those at the bottom of organizations. On the other hand, there is also a great potential for improvement of SHW[9] (see Section 6.6).

Partly due to modern ICT and the intensification of work, a lot of jobs can be done everywhere, and the border between work and private life has blurred, which implies both opportunities and threats to health, wellbeing at work and family life.

Office automation affects worker–manager relationships in a number of ways. It allows middle-level employees a means of providing company executives with reports of production, costs and inventory. This removes the dependence on a few subordinates who traditionally supplied such information. Automation and the use of sensors create ways to monitor each worker's efficiency: through computerized information, managers can, for example, count the number of times per hour that a typist strikes a letter on the keyboard. Managers can also ascertain the number, times and nature of a worker's telephone calls, monitor e-mail, visits to the toilet, etc. The new trend to use people analytics may introduce vicious processes in organizational life.[10] But people monitoring by the use of information technology can also 'help to prevent human errors and facilitate rapid responses to adverse events'.[11] The design of technological systems, including software, also has a great influence on the frequency and nature of 'human error'. Human error is in many cases a result of 'design errors'.[12]

2.2.6 Technology–Organization Interactions

Information technology has many implications for the design and functioning of organizations. The work organization is influenced by IT, making many tasks simpler, creating opportunities for more challenging tasks. Robots and artificial intelligence are increasingly substituting humans. Human–machine collaboration and machine learning are rapidly developing. Telework is blurring the boundaries in time and space between home and work, between work and private life. Due to rapid technological developments and organizational changes, life-long employment is disappearing steadily, while job qualifications are increasingly volatile.

ICT implies a range of (business) opportunities, such as fast responses to emerging events, dynamic risk assessments, identifying complex patterns by analysing big data, etc. But it implies also a range of threats: increase of control over workers, reduction of autonomy, strengthening individualization (growing loneliness – less experience of connectedness), stronger division of labour, etc.

2.2.7 Organization–People Interactions

Careers are increasingly characterized by frequent changes and transitions between various industries. This may imply less loyalty to organizations on the part of the employees and less interest of organizations to invest in new abilities of its personnel.

Due to migration, emancipation and longer life expectancy, diversity in terms of gender, race, religion, age, etc. is growing in many societies and organizations.[13] For companies, diversity implies that it becomes easier to take into account a variety of perspectives in decision-making and policy development. This is increasingly recognized as a business resource. At the same time, it implies challenges in terms of non-discrimination, and it implies that the traditional dominant perspectives are challenged by newcomers – which is often not appreciated. It therefore also implies challenges for the processes and qualities of teams such as communication, cooperation and leadership.

A focus on core activities combined with outsourcing of non-core activities has also triggered a sharp increase of contracting and subcontracting, and of working with all kinds of temporary workers – some of which are formally self-employed. These developments in the human subsystem lead to an increase of precarious jobs with poor SHW, weaken the social bonding and reduce the social capital within organizations. It undermines the responsibility of organizations for SHW and makes it less attractive for companies to invest time, money and effort to get a trusted relationship with an important part of the workforce.

To cope with all changes, capacities for change, adaptation and learning are increasingly important for organizations and people. This has implications not only for educational systems but also for the design of the work organization (opportunities for learning on the job) and the workplace. Both the design of the work organization and the skills and behaviour of people at work are important for a good balance between creativity and control, between autonomy and team work, between being goal-directed and flexibility, between taking action and critical thinking, and between accountability and trust.

2.2.8 Socio-Technical Systems, Continual Change and SHW

The world is full of change, and this has many consequences for organizations (socio-technical systems), work and SHW. The dynamics of the organizational environment and of socio-technical systems strongly influence the SHW challenges and the way organizations and SHW are led and managed. First of all, it is important to think in terms of systems, rather than focus on human behaviour (e.g., human error). It is important to address the SHW challenges/problems not in silos and not always through solving upcoming problems one by one. The dynamics of socio-technical system imply that problems/challenges in one part of the system may impact other parts of the system, and vice versa. Changing one factor or process may impact several others as well.

It is more important to think in terms of dynamic processes of interaction and change, and less important to think in terms of stable states. As a consequence, many concepts and mental models that were useful in the past have lost a great part of their relevance. Vision Zero implies that the vision needs to depart from a dynamic and systemic worldview that does justice to the continual changes in organizations and the SHW risks associated with these dynamics. A variety of perspectives, representing at least the technological, organizational and human view on socio-technical systems and perhaps external stakeholders, are complementary. These imply a strong need for multi-disciplinary cooperation and dialogues wherein implicit assumptions are made explicit and mutually challenged.

The dynamics of change combined with the ambitions of Vision Zero require leadership, proactivity, anticipation and learning. This strengthens the capability to adapt and innovate and the resilience of the socio-technical system. The potential of IT should thereby be used combined with the participation of the workers/users. Through dynamic real-time risk assessments and automated corrective actions, modern ICT can improve prevention in many aspects of SHW. It may lead to a new dynamic balance between people and technology for ensuring SHW.

2.3 SAFETY, HEALTH AND WELLBEING IN A WORLD OF CHANGE

In the former section, we have seen how changes in organizations, considered as socio-technical systems, influence work and indirectly SHW.

Safety is attributed to systems, things, conditions, processes, etc., not to individual people. When people are safe, the meaning is that the systems and processes at hand are safe for them. When a product is regarded safe, it implies that the probability of an incident with the product in combination with the potential effect thereof is regarded as minimal or acceptable. Engineers need standards for acceptable safety hazards and risks, otherwise, they cannot develop safe systems, products, processes, etc.

Different from safety, health and wellbeing are states of individual people. Only in a metaphorical sense, we can speak about healthy products or healthy conditions – meaning they are good for – or at least do no harm to – the health and wellbeing of individuals. Safe work or healthy work means that the working conditions are good for safety or health (having a positive influence) and that the people at work are adequately protected against the negatives (SHW hazards and risks).

The dynamics of change have generated new types of (emerging) SHW risks and triggered (1) greater attention to SHW management as a dynamic *process*, (2) new perspectives on SHW that better suit today's challenges and (3) more dynamic concepts of SHW.

A key concept for SHW is *risk*. Risk refers to the possibility of injury, illness, disease or loss. It is an effect of uncertainty about the effects and implications of an activity, process or situation. Usually, it refers to undesirable and unexpected consequences of activities. Traditionally a risk is defined as a combination of the consequences of an event and the associated likelihood of occurrence. Usually, several risks are relevant simultaneously which may develop through a variety of scenarios, each with specific likelihoods of occurrence and type or seriousness of consequences.

The ISO 31000 standard on risk management[14] defines risk as *the effect of uncertainty on objectives*. The definition implies that a risk cannot only be a negative deviation from the objectives (downside risk) but also be a positive deviation (upside risk). The definition emphasizes that we should not only focus on negative surprises (as most people understand risk) but also on the positive surprises, which imply an opportunity to improve. The ISO 31000 definition is, however, less useful for proactively managing SHW and pursuing Vision Zero. When (SHW) risks are a kind of side effect of business objectives, it implies they are a second-order priority for businesses. When positive deviations are understood as a kind of SHW risks, not as opportunities for improvement, that may easily make business leaders consciously accept significant SHW risks (including the potential negative consequences).

When a risk materializes that is never part of a preconceived plan, only plans to reduce risks are useful. A risk is always resulting from unintended consequences, even when a certain risk is consciously taken. The materialization of a risk is therefore always a kind of surprise. Though you can be prepared for surprises, they are not planned. Aims and planning for a specific number of surprises (e.g., lost time incidents) are therefore never SMART. It involves a chance process with a statistical outcome.

The term *emerging risks*, frequently used in our area, refers to phenomena that are relatively new compared to traditional SHW risks. Especially psychosocial risks have developed into major challenges. Psychosocial risks have rather complex backgrounds, but there is a lot of knowledge about them available, and they sure can be prevented.

The range of challenges implied by psychosocial hazards and risks is illustrated in Table 2.1. The (re)design of the work organization is thereby key for prevention.

TABLE 2.1
Healthy and unhealthy psychosocial working conditions

Dimensions of Work Organization	Unhealthy Psychosocial Working Conditions (Psychosocial Hazards)	Healthy Psychosocial Working Conditions
Purpose	Lack of definition of, or agreement on, organizational objectives and absence of human or social business values	Clear organizational objectives, alignment of worker and organizational values
Job content	Lack of variety or short work cycles, fragmented or meaningless work, under use of skills, high uncertainty, continuous exposure to people through work, social or physical isolation	Meaningful work, appropriate use of skills, work retaining employee interest and engagement, appropriate support
Workload and work pace	Work overload or underload, machine and algorithm pacing, high levels of time pressure, continually subject to deadlines	Appropriate level of workload, appropriate work pace, human in control work pacing, sensible and achievable deadlines
Work schedule	Shift working (especially irregular), night shifts, inflexible work schedules, unpredictable hours, long or unsociable hours	Sensible shifts and reasonable working hours to maintain work–life balance, flexible working practices
Control	Low participation in decision-making, lack of control over workload, pacing, shift working	Participation in decision-making, control at work
Environment and equipment	Inadequate equipment availability, suitability or maintenance; poor environmental conditions such as lack of space, fresh air, poor lighting, excessive noise	Good physical working conditions
Role in organization	Role ambiguity, role conflict	Clear roles and responsibilities
Home–work interface	Conflicting demands of work and home; issues arising by telework from home	Good work–life balance

Source: Adapted from Leka, S., Jain, A., Lerouge, I., (2017) Work-related psychosocial risks: Key definitions and an overview of the policy context in Europe, in Lerouge, I., (ed), *Psychosocial Risks in Labour and Social Security Law.* Dordrecht: Springer Science. pp. 15–30.

2.3.1 SHW Management as a Dynamic Process

Dynamic management processes are necessary in order to cope with ongoing change. Since the end of the twentieth century, a lot of efforts have been directed at the development, implementation and certification of OSH management systems.[15] Periodic audits and management reviews ensure that adaptation to new circumstances is regularly considered. OSH management systems combine a system view with a dynamic process, two factors that explain its successes. However, OSH management systems have regularly led to too much focus on paperwork, prescriptions and compliance with standards while the intended process of continual improvement has often not materialized (see Chapter 8).

Another model for a dynamic approach is the World Health Organization's model for healthy workplaces[16] (see Figure 2.2).

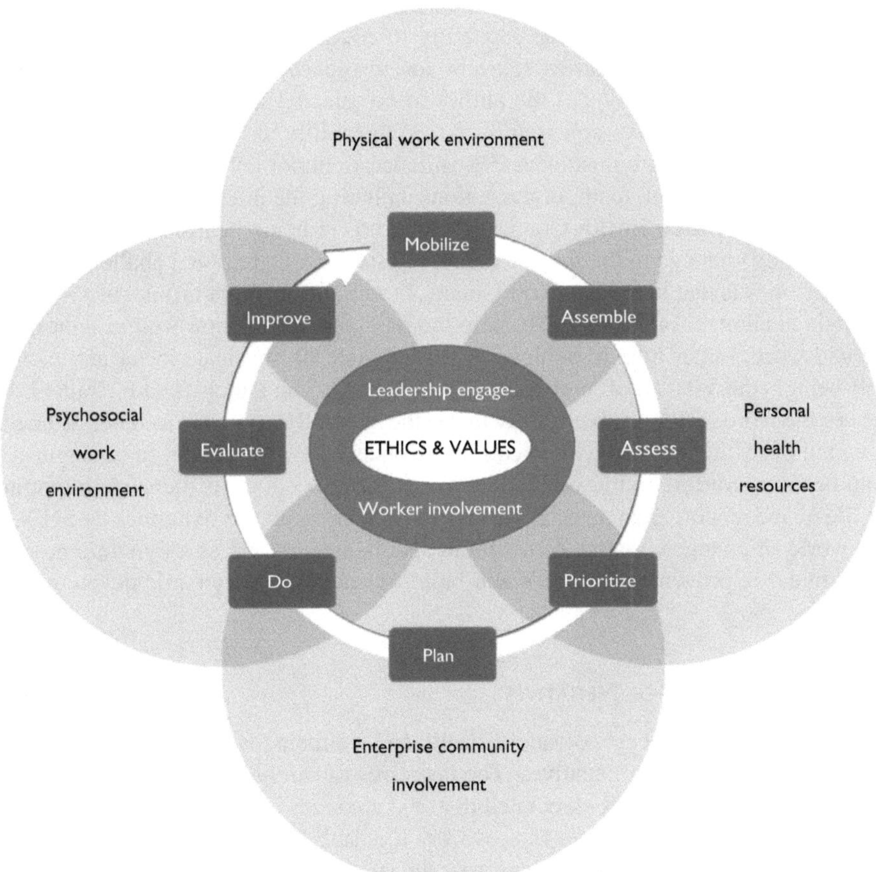

FIGURE 2.2 The WHO Model for health work places. (WHO, (2010) Healthy Workplaces: A Model for Action. Geneva: World Health Organization.)

Different from the ISO models, the WHO model has *ethics* and *values* at the core of its model. Ethics and values are very relevant for Vision Zero (see Chapter 4). Leader engagement and worker involvement are seen as key. The WHO model[17] distinguishes between four areas for improvement: the physical work environment, the psychosocial work environment, personal health resources and enterprise community involvement.

Digitalization is now the major step forward in making SHW management more dynamic. Using sensors and cameras, it is in an increasing number of cases possible to assess and control risks in real time. See Section 6.6 for its preventive potential.

2.3.2 Resilience

The concept of *resilience* originally stems from the area of ecology. Today it is used in a broad variety of fields, including health, safety, social work and family studies, organizational studies, economics and politics. An important quality of the concept of resilience is its dynamic nature. Resilience can be defined as the ability to succeed under varying conditions and the capability to cope with new challenges. Schelvis et al. distinguish three abilities relevant for resilience: (1) the ability to change and adapt, when necessary, (2) the ability to be 'elastic' and recover quickly from changes, adverse events or constraints and (3) the ability to be and remain confident in and after difficult circumstances.[18] Resilience is important at different organizational levels (individual, team, organization) including the interaction thereof.

In 2011, the Dutch Health Council proposed to redefine health as 'the ability to adapt and self-manage in the face of social, physical, and emotional challenges'. The main reason was that in today's world, many people suffer from chronic diseases, and the main challenge for doctors is to help them be able to function well in their daily life and work, rather than to achieve 'a perfect state of physical, social and mental well-being' (the WHO's official definition of health). The Dutch Health Council did not use the word resilience but is easy to see they actually proposed to redefine health as a kind of resilience. Today they call their concept *positive health*, to distinguish it from health according to the WHO definition. The concept of resilience is important for SHW and Vision Zero as it is relevant for dealing with the dynamics of SHW in 'the world of change'. Vision Zero should also be understood as a dynamic process of continual improvement in SHW and build on each of the dynamic developments mentioned above.

2.3.3 The Absence of Negatives

Though each of the concepts of safety, health and wellbeing is positive, they are often defined as the absence of negatives. The negatives of safety are incidents, accidents, injuries, disasters as well as associated material damages and economic losses. Their absence suggests that it is safe. E.g., we say it is safe to cross a street when we are confident that during the time that we cross the street, there are no conditions that will cause an accident to happen. Regarding health and wellbeing, the negatives are (physical or mental) health complaints, illnesses, diseases and disorders, injuries, reduced (work)abilities, associated with sickness absence and reduced productivity, etc.

It is much easier to identify and focus on the negatives and to try to prevent, reduce or mitigate them than trying to identify, develop or maintain a state of 'being safe and healthy'. This is why in safety the aim is often to reduce the frequency or severity of accidents and incidents. Regarding health, the aim is often to prevent, cure or mitigate diseases and disorders. As a result the absence of the negatives is often taken for the positive concept of SHW themselves. SHW are then understood as the condition of being protected from (SHW) harm or other danger, or as states with a low and accepted level of negatives.

The focus in safety is usually on acute negative events, often with immediate effect, though acute events can also have long-term impacts (e.g., death, infirmity, post-traumatic stress) and factors that contribute to accident causation may exist over significant periods of time. The focus in health and wellbeing is usually on chronic exposures and their long-term impact, though acute events like violence, mobbing in the workplace or exposure to extreme levels of noise or toxic agents are quite relevant for health and wellbeing. In sectors like health care and police, repeated exposure to events of violence is increasingly relevant.

The focus on negatives has important limitations: it is only in hindsight that you can count accidents and incidents, while diseases and disorders can only be identified when they are already occurring – often after a (long) incubation period. The absence of negatives reflects SHW in the past. Does the absence of work-related incidents in the last week imply that it will be safe at work in the coming week, or month? Does the absence of work-related diseases during the last month imply such diseases will not occur in the coming month or years? Unfortunately, the answer to such questions is no: The absence of negatives has little predictive value for SHW in the future. Though the term Vision Zero may suggest otherwise, it is therefore vital for Vision Zero not to focus only on 'lagging indicators' that reflect negative past events but to focus on proactive leading indicators (see Section 9.2).

People who criticize Vision Zero often emphasize that it refers to zero negatives. But Vision Zero is a vision, an ambition and is future-directed. The ambition that work and production should be free of work-related accidents and diseases is an inspiring vision to everybody who has personally experienced a work-related accident or illness or has witnessed the impacts thereof in close colleagues, friends or relatives. Loss aversion is the term for the fact that most people's responses to losses are much more powerful, psychologically, than to gains.[19] Loss aversion implies that Vision Zero is much more appealing than it would be when it only expressed the ambition of excellence in SHW.

The frequency of incidents and diseases also has a statistical dimension: their occurrence varies in time implying uncertainties in when and where they will occur. In large organizations, the statistical variations will usually middle out, but when such frequencies are applied in SMEs, they are by definition unreliable. When the frequencies in large companies are very low (zero or approaching zero), there will always be notable variations (e.g., in lost time injury frequencies, fatal accidents, sickness absence) over time that do not necessarily reflect safety or health adequately. After an unusual poor SHW performance, everything else remaining the same, an improvement in performance can be expected solely for statistical reasons (known as regression to the mean). That statistical process makes it also difficult to assess whether an improvement in SHW performance was due to the intervention or the statistical nature of unplanned events.

2.3.4 Positive Approaches to SHW

All three aspects, safety, health and wellbeing, are in itself positive concepts. Safety refers to being 'safe'. It is a state wherein the integrity of life, health (and properties), is ensured. Safety at work is then the condition or state that life and health are ensured at work, especially in relation to potential sudden and accidental events. *Health*, as defined by the WHO, is *a state of complete physical, mental and social wellbeing, not merely the absence of disease or infirmity*.[20] In this definition of health, wellbeing is already included; in this book, we will use the concept of 'wellbeing' as equal to positive mental health. It is important to keep in mind that SHW are dynamic positives and always interact with the (work) organization and other aspects of the work environment and that opportunities for improvements imply a need for forward thinking.

Antonovsky[21] was the first to assume that there are not only negative factors but also positive factors that generate health and wellbeing. He named them *salutogenic* (health-generating) factors. Antonowsky's work formed the start of positive psychology which aims to promote the wellbeing of people, individually and collectively, by strengthening positive factors and events that contribute to wellbeing. In the positive psychology, positive terms are used such as autonomy, coherence, engagement, flourishing, flow, happiness, love, meaning, personal growth, positive relationships, purpose and self-esteem. Such positive concepts contribute to the improvement of health and wellbeing. There is also a growing attention to the positive influence that work can have on health and wellbeing. Traditionally, work is regarded as a potential threat to SHW. There is, however, good evidence that it is often much better for health and wellbeing to be employed than to be unemployed.[22] This depends on the 'quality of work'; a good quality of work has a positive impact on health and wellbeing.[23] Research on return to work has also highlighted several examples where work contributes to rapid recovery after illnesses and injuries.

A positive concept related to safety is reliability, which is often used in reference to high-reliability organizations (HROs[24]). For HROs, it is essential to stay attentive to unexpected and upcoming risks which require among others a culture of organizational mindfulness. The concept of Safety-II[25] is a positive concept. The main idea of Safety-II is to focus on 'what goes right' instead of a focus on 'what goes wrong' (Safety-I). In Safety-II, it is rightly emphasized that development and adaptation are essential for the long-term survival of organizations as well as people. Work processes are thereby regarded more important than organizational structure. In Safety-II, people are not seen as a source of potential errors, but as an asset, an important buffer of performance variability in social-technical systems. In Safety-II, attention to 'negatives' such as accidents, near misses or risks is regarded as reactive and inferior to attention to 'what goes right'. Managing performance variability implies taking counter-measures in case of small deviations from the desired situation in the dynamic work process – in this respect it can be regarded as reactive as well. Using performance variability to improve safety can be regarded as an extension

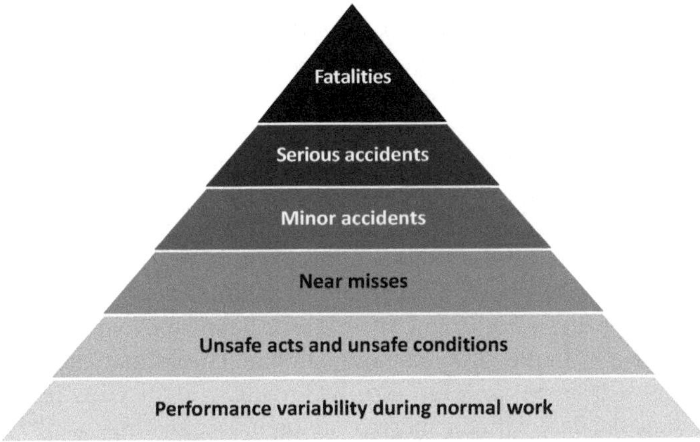

FIGURE 2.3 The extended Heinrich pyramid including performance variability.

of the well-known Heinrich pyramid (see Figure 2.3). The basis of the pyramid is thereby extended with the attention to performance variability during normal work.

As in the normal Heinrich pyramid, it is assumed that controlling the activities on the basis of the pyramid will automatically result in controlling the events at the top of the pyramid, i.e., serious accidents and fatalities. However, today it is well-known that this is certainly not always the case. That depends on the scenarios relevant for the various events, and whether the causal factors that play a role at the bottom of the pyramid also play a role in the scenarios causing the more severe events.[26] The statement that serious accidents have the same causes as performance variability during normal work is therefore misleading, as it is only rather exceptionally the case. As Safety-II focuses on work processes (i.e., in the stage of production activities), it is not a very useful concept in the design stage.[27]

THE SAFETY-II APPROACH AND LEARNING FROM EXCEPTIONAL VARIATION

The Safety-II approach focuses on learning from normal variation in work and production processes and on attention to positive variation, which may imply opportunities for improvement that could easily be overlooked if there is only attention to negative occurrences like incidents.

Different from what Hollnagel and other promotors of Safety-II claim, learning from normal variability is often not adequate for coping with unexpected events that are beyond the normal variability. Learning from normal variation requires first-order (adaptive) learning. Coping with factors beyond normal variability requires more fundamental (innovative) learning processes.

The most prominent example was the struggle about what to do when the world was suddenly confronted with the Covid-19 pandemic. There was a school of thought that believed that it could be handled best by using the experience with common infectious diseases stemming from viruses like influenza (= the experience from normal variation). Unfortunately, the resilience of the human body proved to be insufficient for dealing with Covid-19 which led to the decease of millions of mostly vulnerable people. After some time, good vaccines became available that were a form of classical prevention (comparable with Safety-I). Interestingly, when many vulnerable people were protected by the vaccines, and many other vulnerable people died or survived the disease, the situation changed again. Survivors of the disease and people protected by the vaccines have developed a considerable resistance against Covid-19. In the meantime, the dominant Covid-19 viruses had mutated to the less malicious Omicron variation. In the new situation with populations that had gained a lot of resistance, it became useful to deal with the Covid-19 virus in ways very similar to how we deal with the more common influenza virus.

The case nicely illustrates that Safety-I and Safety-II measures are often complementary and that Safety-II can become more relevant in situations where Safety-I is already well-developed.

In Safety-II, the human qualities of flexibility, creativity and learning are rightly seen as essential for dealing with the uncertainties associated with processes of emergence (thereby carefully avoiding the term emerging risks – risks are associated with Safety-I). Adaptation, competence, flexibility, learning and so on were always regarded as relevant to safety. But Safety-II has certainly generated greater attention for opportunities for improvement and for how work is actually done. The renewed attention to how work is actually done is also emphasized in what is now called 'the new view on safety' or 'safety differently' which can be regarded as a correction of the too-great emphasis on the prevention of 'human error' and behavioural control in several more traditional safety approaches. In this 'new view', there is a positive appraisal of craftmanship and workmen's' knowledge about practical issues confronted during normal work.

In Safety-II approaches, no attention is given to negatives like risks or 'what goes wrong'. But according to Popper,[28] the focus on what goes wrong is key to scientific progress. Science progresses when a theory is shown to be wrong and a new theory is introduced that explains the phenomena better, not by focusing on what goes right or according to expectations. In this respect, it is rather paradoxical that Hollnagel claims Safety-II to be the future of safety. Finally, there is also empirically driven critique on Safety-II and the associated 'new view' on safety, because there is a lack of evidence that it leads to a better safety performance.[29]

2.3.5 SHW as Resources for Work and Productivity

Another positive approach to SHW is to acknowledge them as business resources that are key to business or economic performance. For organizations, it is attractive to invest in resources or capabilities that positively impact its functioning. Health and wellbeing can be regarded as factors that codetermine the functioning of people (human and social capital) and therefore as production factors.[30] For instance, good mental health is a resource for creativity and innovation while mental health problems are closely associated with productivity losses (due to sickness absence or presenteeism). The resource value of health and wellbeing is further illustrated in Table 2.2 which builds on the formal WHO definition of health as a complete state of physical, mental and social wellbeing, not merely the absence of disease or infirmity.[31] Safety is regarded by the WHO as a (health) matter of *injury prevention*.

Besides being a business resource, physical and mental health represent economic value to individuals in terms of the capability to generate income. Good health is important to remain economically independent from other people and social security bodies.[32] Safety can also be regarded as a business resource, as safe work processes imply that the processes are under control while disturbances are kept to a minimum. Accidents and incidents generate business losses that can be avoided. Safety excellence and striving to achieve zero accidents is therefore regarded as a key dimension of world-class manufacturing[33] (see Section 10.1). Of course, injuries also affect the physical and mental health of people, and in this way indirectly the functioning of people in production.

TABLE 2.2
Health and wellbeing aspects and their relationships with productivity and economic benefits

Health as a Resource	Physical Health	Mental Health	Social Health	Total Health
Power	*Muscle power*	*Brain power*	*Social and communicative power*	*Man/human power*
Typical work	*Handling materials*	*Processing information*	*Networking, managing*	*Creating and developing*
Resource theory	*Classical human resource*	*Intellectual capital*	*Social capital*	*Human and social capital*
Typical technologies – competing and/or enabling	*Traditional technologies*	*Information technology*	*Communication and transport technology*	*Both classical technology and information and communication technology (ICT)*
Economies where this resource is most relevant	*Industrial society*	*Knowledge economy*	*Network society, global economy*	*Sustainable economy*

Source: Elaborated on Zwetsloot, G.I.J.M., van Scheppingen, A.R., (2007) Towards a strategic business case for health management, in Johansson, U., Ahonen, G., Roslander, R., (eds), *Work Health and Management Control*. Stockholm: Thomson Fakta. pp 183–213.

2.3.6 Collaborative Safety and Health

In Japan, the concept of Safety 2.0 is developed and promoted, which is called *collaborative safety* in Western countries to avoid confusion with Safety-II. Collaborative safety makes use of the rapidly increasing possibilities of Information Technology, including Artificial Intelligence (AI).[34] By combining sensors with IT, dynamic, real-time risk assessments can be made, for instance, with respect to a collision or the danger of getting stuck. In the same way, it can help to prevent work-related illnesses due to exposure to toxic materials, noise, vibrations, etc. beyond the threshold limit value or lower level resulting from working according to best practices. Real-time information on safety and health risks can be given to the workers and their supervisors, but the feedback can also be used to adapt the functioning of *robots*. The so-called *collaborative robots* or cobots can adapt their functioning at any moment, e.g., after an unexpected movement of the operator. The information can be fed back to an IT platform supervising several workers and cobots. The level of qualification and experience of the worker can be communicated via a chip (e.g., in the helmet or clothes) so that the cobots 'know' the worker and can adapt the level of speed and type of movements according to their abilities. Advanced sensors can notice that a worker is not functioning as expected, e.g., because he or she is distracted or tired, and give relevant feedback to adapt the functioning of the cobot.[35] Though the real-time monitoring of the behaviour of the worker could be experienced as a nuisance, the Japanese experience is that collaborative safety increases the wellbeing of the worker as the discomfort of 'constantly feeling uneasy about potential safety risks' is eliminated this way. In this way, the responsibility for working safely is shared between the intelligence of the worker and the intelligence of the technological device. While in Safety-II workers are regarded as the key to managing performance variability, in collaborative safety this function is mostly built into the technological parts of the socio-technical system.

The development of collaborative safety raises several new ethical questions about the responsibility for the functioning of the technology and the safety of the socio-technical system as a whole. When the feedback via the technological system fails, is that the responsibility of the worker? Of the supervisor? Of the engineer who developed the system? Of the software engineer who programmed the IT? Of the technical service responsible for the maintenance of the technical system and the updates of the software?

As the development of information technology is going faster than ever (5G, 6G, Digital twins, AI, AI on a chip), it can be expected that collaborative safety will soon be applied in almost every sector, though the long-term impacts are difficult to oversee. The flip sides of the rapid IT development are that the safety and reliability of software are very difficult to assess and that cyber security risks often also imply safety and privacy (wellbeing) risks. Software can introduce new risks which are difficult to assess and often depend on the interactions with the context where it is used. This is especially relevant in high-risk computerized systems.[36] Any technology can be used for better or for worse. The negative sides of the IT developments for SHW are most significant in platform work and the gig economy. To use IT mostly for its positive impacts will require that the development of IT and its applications is combined with a well-developed ethical compass and explicit attention to the well-being of the workers.

2.3.7 SHW as Human and Social Values

Safety, health and wellbeing are more than merely rational concepts: they are also basic human and social values. They have an ethical or moral dimension which is important for each and every individual, which should also be important for any profit and non-profit organization. The value dimension of SHW is an important driver behind the adoption of Vision Zero. Markku Aaltonen rightly stated that 'zero accidents is the only ethically sustainable long-term goal for safety management'. Similarly, it can be said that the goal of zero work-related (physical or mental) illnesses is the only ethically sustainable goal for health and wellbeing at work.

Not each and every individual is aware of SHW as values and acts accordingly. As long as people feel well, they can easily ignore SHW; negative experiences (accidents, injuries, illnesses) often trigger the acknowledgement of SHW as important values. This is true for individuals (workers and managers) as well as organizations and institutions. Acknowledging SHW as human and social values will impact the decisions and policies of organizations and the acting and thinking of individuals. As values, they are important for deciding what is the 'right' thing to do. See Section 4.4.

When SHW is seen from the value perspective, it is easy to see that the three concepts have much in common. Then, they have more commonalities than differences. The value dimension is therefore an important driver for more integrated approaches of SHW. A very important example of such an integrated approach is the Total Worker Health© concept and approach, developed by the National Institute for Occupational Safety and Health (NIOSH) in the United States. Total Worker Health© is defined as 'policies, programs, and practices that integrate protection from work-related safety and health hazards with promotion of injury and illness-prevention efforts to advance worker well-being'.[37] The ethical-value aspect of SHW also means that SHW is an important aspect of business ethics and corporate social responsibility.[38]

Closely related to the value aspect of SHW is *the rights aspect*, which was already addressed in Section 1.1. In 2022, the ILO Governing Body added the 'right to a safe and healthy working environment' to the ILO framework on fundamental principles and rights at work. As the UN organization involved in issues of work, it is governed by a tripartite board (employers, trade unions and governments). Consequently, this is an almost universal recognition by vital stakeholders. Note that the right concerns 'a safe and healthy working environment', not the right of individuals at work to be safe and healthy. The ILO also included the fundamental right to a safe and healthy working environment as part of its declaration of *social justice* for *fair globalization*. The same value aspects are associated with the concept of 'decent work' which has been promoted for decades by ILO as a key concept.[39]

While a system approach to SHW implies boundaries of the system in terms of space and time, this is not the case for SHW as human or social values or as a fundamental right. Values and fundamental rights are largely valid independent of space, time or appearance. Consequently, the distinction between occupational safety and other forms of safety (public safety, product safety, commuter safety, etc.) is less relevant. The same is true for physical and mental health aspects of occupational health, environmental health and public health. From an ethical or value perspective, the responsibility for good SHW does not end at the company fence and also implies certain broader responsibilities for SHW of business partners, customers and neighbours (see Section 10.3).

When companies and institutions acknowledge that safety, health and wellbeing at work are three closely related basic human and social values, these values will give direction to the organizational strategy as well as to the development of the organizational culture. The value dimension of SHW is therefore an important fundament for a proactive prevention strategy and for the development of a prevention culture. It is essential for Vision Zero.

2.4 DEALING ADEQUATELY WITH CHANGE

As we have seen, change is a constant factor in today's organizations. It is important to realize that the best practices of today are perhaps no longer the best practices of tomorrow; they can become even obsolete in a few years' time. It is therefore important to deal adequately with change. It is thereby important not to 'run behind the facts' but to be somehow prepared for change.

This implies that the capacity for dealing with change and emerging problems requires specific attention; that capacity needs to be increased. Anticipating change is better than reacting to change. This requires learning, from leaders, the workforce, and the organization as a whole. A quote from Peter Allen illustrates the importance of individual and organizational learning in a world full of change. 'In *the shifting landscape of a world in continuous evolution, the ability to climb is perhaps what counts, and what we see as a result of evolution are not species of firms with "optimal behaviour" at each instant, but rather actors that can learn!*'.[40] The topic of individual and organizational learning is therefore quite relevant for Vision Zero. It will be addressed further in Chapter 4.

2.5 CONCLUSIONS

Continual changes in economy, technology, demography, ecology and culture have many implications for organizations, work and SHW. In a world of change, the only constant factor is change. Continuous change implies increasing uncertainties for organizations. Organizations have to adapt to the changes in their organizational environment to ensure their continuity as well as SHW.

Approaches to SHW that are implicitly based on a static worldview are less useful in the dynamic world of work. Organizations need to cope with the dynamics of change and novel circumstances generated by innovations and societal changes. As change frequently influences the interactions between the (technological, organizational and human) subsystems in organizations, a system approach is increasingly relevant for managing SHW risks. Focussing on one factor only (e.g., technology or human error) is no longer adequate in a dynamic context.

Change impacts the nature of hazards and risks, positively or negatively. It triggered the development of new alternative views on SHW and risk management. While each of the three dimensions of SHW is in itself positive, SHW is often operationalized in terms of the absence of negatives (accidents, illnesses, etc.). In the last decades, positive approaches have come to the forefront, such as positive psychology, Safety-II, SHW as business resources and SHW as human and social values.

Change has triggered the development of dynamic concepts like *resilience* and *positive health* which better suit the dynamics of change. The emphasis is no longer only on the protection of the workers; instead, they are given more active roles and more responsibilities. The practical wisdom of workers and the added value of participation is thereby acknowledged as indispensable for dealing adequately with change.

Each of the new approaches has its strengths and limitations. It is necessary to acknowledge their complementarities and potential synergies. There is a great need for a new dynamic and integrated approach to SHW prevention and promotion which is conducive to strategy.

Based on the acknowledgement of safety, health and wellbeing at work as basic human and social values and the recognition of a healthy and safe work environment as a fundamental right, Vision Zero is that integrated approach. The Vision Zero approach combines a focus on systemic prevention of negative impacts while promoting positive achievements. Vision Zero builds on the common aspects of SHW. It makes use of synergistic potentials, rather than emphasizing differences or continuing silos. In this way, Vision Zero can function as an umbrella that makes use of all available knowledge and approaches to generate a long-term process of continual improvement in SHW performance.

NOTES

1 Yates, J., (1989) *Control through Communication: The Rise of System in American Management*, Baltimore: John Hopkins University press.

2 ILO, (2024) *Ensuring Safety and Health in a Changing Climate*. Geneva: International Labour Organization.

3 Beck, U., (1992) *Risk Society. Towards a New Modernity.* London: Sage.

4 For instance, Hollnagel, E., (2014) *Safety I and Safety II. The Past and Future of Safety Management*, Farnham: Ashgate; and Hohnen, P., Hasle, P., Jespersen, A.H., Madsen, C.U., (2014) Hard work in soft regulation: A discussion of the social mechanisms in OHS management standards and possible dilemmas in the regulation of psychosocial work environment. *Nordic Journal of Working Life studies*, 4 (3), 13–30. Doi: 10.19154/njwls.v4i3.4177.

5 Leveson, N., (2020) Are you sure your software will not kill anyone? *Communication of the Association for Computing Machinery*, 20 (2), 25–28. Doi: 10.1145/3376127.

6 Morgan, G., (1986) *Images of Organization.* London: Sage.

7 Pasmore, W., Winby, S., Mohrman, S.A., Vanasse, R., (2019) Reflections: Sociotechnical systems design and organization change. *Journal of Change Management*, 19 (2), 67–85. Börnfelt, P.O., (2023) The Sociotechnical school. *Work Organisation in Practice*. Palgrave Macmillan, Cham. Doi: 10.1007/978-3-031-21667-1_.

8 Crawston, W., Malone, T.W., (1988) Information technology and work organization. Chapter 49. *Handbook of Human–Computer Interaction.* Cambridge: MIT Press. pp. 1051–1070.

9 IEC, (2020) *White Paper: Safety in the Future*. Geneva: International Electrotechnical Commission. www.iec.ch/basecamp/safety-future-0 (Accessed 16 December 2024).

10 Gal, U., Blegind Jensen, T., Stein, M.K., (2020) Breaking the vicious cycle of algorithmic management: A virtue ethics approach to people analytics. *Information and Organization*, 30 (2), 100301. Doi: 10.1016/j.infoandorg.2020.100301.

11 Bates, D.W., Gawande, A.A., (2003) Improving safety with information technology, *New England Journal of Medicine* 348, 2526–2534. Doi: 10.1056/NEJMsa020847.

12 Norman, D.A., (1983) Design rules based on analyses of human error. *Communications of the Association of Computing Machinery*, 26 (4), 254–258.

13 Jain, A., Leka, S., Zwetsloot, G.I.J.M., (2018) Work, health, safety and well-being: Current state of the art. In: Jain, A., Leka, S., Zwetsloot, G.I.J.M., (eds), *Managing Health, Safety and Wellbeing – Ethics, Responsibility and Sustainability.* Dordrecht: Springer Science. pp. 1–32.

14 ISO, (2018). *ISO 31000:2018, Risk Management – Guidelines*. Geneva International standardisation Organization.

15 Zwetsloot G.I.J.M., (2000) Developments and debates on OHSM system standardisation and certification, in Frick, K., Quinlan, M., Jensen, P.L., Wilthagen, T., (eds), *Systematic Occupational Safety & Health Management- Perspectives on an International Developmen*t. Oxford: Pergamon-Elsevier Science, pp. 391–412.

16 WHO, (2010) *Healthy Workplaces: A Model for Action*. Geneva: World Health Organization.

17 The value aspects, leadership and participation are also important elements for developing 'organizational health'. See Lowe, G., (2010) *Creating Healthy Organizations.* Toronto: University of Toronto Press.

18 Schelvis, R.M.C., Zwetsloot, G.I.J.M., Bos, E.H., Wiezer, N.M., (2014) Exploring teacher and school resilience as a new perspective to solve persistent problems in the educational sector – A case of the Netherlands, *Teachers and Teaching, Theory and Practice,* 20 (5), 622–637. Doi: 10.1080/13540602.2014.937962.

19 Kahneman, D., Tversky, A., (1992) Advances in prospect theory: Cumulative representation of uncertainty. *Journal of Risk and Uncertainty*, 5 (4), 297–323. doi:10.1007/BF00122574.

20 WHO, (1948) *Constitution*. Geneva: World Health Organization.

21 Antonowsky, A., (1987) *Unravelling the Mystery of Health; How People Manage Stress and Stay Well.* San Francisco: Jossey-Bass.

22 Waddell, G., Burton, A.K., (2006) *Is Work Good for Your Health*? London: The stationary Office.

23 Taylor, M., Marsh, G., Nicol, D., Broadbent, P., (2017) *Good Work. The Taylor Review of Modern Working Practices*. London: Department for Business, Energy, and Industrial Strategy.

24 Weick, K.E., Sutcliffe, K.M., (2007) *Managing the Unexpected: Resilient Performance in an Age of Uncertainty* (2nd edition). San Francisco, CA: Wiley.

25 Hollnagel, E., (2014) *Safety I and Safety II. The Past and Future of Safety Management.* Farnham: Ashgate.

26 Bellamy, L.J., (2015) Exploring the relationship between major hazard, fatal and non-fatal accidents through outcomes and causes. *Safety Science,* 71, Part B, 93–103. Doi: 10.1016/j.ssci.2014.02.009.

27 Leveson, N., (2020) Safety *III – A Systems Approach to Safety and Resilience*. Cambridge: MIT.

28 Popper, K., (1959) *The Logic of Scientific Discovery*. Hutchinson & Co. (first published in German in 1935 as *Logik der Forschung,* Vienna: Springer).

29 Cooper, D., (2022) The emperor has no clothes: a critique on safety II. *Safety Science*, 152, 105047. Doi: 10.1016/j.ssci.2020.105047.

30 Zwetsloot, G.I.J.M., van Scheppingen, A.R., Dijkman, A., Heinrich, J., den Besten, H., (2010) The organizational benefits of investing in workplace health. *International Journal of Workplace Health Management*, 3(2), 143–159. Doi: 10.1108/17538351011055032.

31 WHO, (1948) *Constitution.* Geneva: World Health Organization.

32 Zwetsloot, G.I.J.M., Pot, F.D., (2004) The business value of health, *Journal of Business Ethics*, 55, 115–124. Doi: 10.1007/s10551-004-1895-9.

33 D'Orazio, L., Messina, R., Schiraldi, M.M., (2020) Industry 4.0 and world class manufacturing integration: 100 technologies for a WCM-I4.0 matrix. *Applied Sciences*, 10, 4942. DOI: 10.3390/app10144942.

34 Mukaidono, M., Takaoka, H., Ogihara, H., Ariyama, M., Fujita, T., (2018) Japan's approach for the realization of the future safety concept by implementing collaborative safety technologies, in: *9th International Conference on Safety of Industrial Automated Systems (SIAS),* Nancy (France),

pp. 77–87. See also IEC (2020) White Paper: *Safety in the Future.* International Electrotechnical Commission, Geneva. www.iec.ch/basecamp/safety-future-0 (Accessed 16 December 2024).

35 Mukaidono, M., Takaoka, H., Ogihara, H., Ariyama, M., Fujita, T., (2018) Japan's approach for the realization of the future safety concept by implementing collaborative safety technologies, in: *9th International Conference on Safety of Industrial Automated Systems (SIAS),* Nancy (France), pp. 77–87. See also IEC (2020) White Paper: *Safety in the Future.* International Electrotechnical Commission, Geneva. www.iec.ch/basecamp/safety-future-0 (Accessed 16 December 2024).

36 Leveson, N., (2020) Are you sure your software will not kill anyone? *Communication of the Association for Computing Machinery*, 20 (2), 25–28. Doi: 10.1145/3376127.

37 NIOSH, (2023) *What Is Total Worker Health©?* www.cdc.gov/niosh/twh/about/?CDC_AAref_Val=https://www.cdc.gov/niosh/twh/totalhealth.html (Accessed 14 December 2024).

38 Jain, A., Leka, S., Zwetsloot, G.I.J.M., (2018) Responsible and ethical business practices and their synergies with health, safety and wellbeing, in Jain, A., Leka, S., Zwetsloot, G.I.J.M., (eds), *Managing Health, Safety and Wellbeing, – Ethics, Responsibility and Sustainability*. Dordrecht: Springer Science, pp. 99–138.

39 Ghai, D., (2008) Decent work, concept and indicators. *International Labour Review*, 42 (2), 113–145.

40 Allen, P.M., (1988) Evolution, innovation and economics, in Dosi, G., Freeman, C., Nelson, R., Silverberg, G., and Soete, L., (eds), *Technical Change and Economic Theory*. London, New York: Pinter. p. 107.

3 Vision Zero

3.1 WHAT IS VISION ZERO?

'Vision Zero' is a transformational approach to prevention that integrates the three aspects of Safety, Health and Wellbeing at work (SHW). Vision Zero is based on the assumption that all (serious) accidents, harm and work-related ill-health are preventable. *Vision Zero is the ambition and commitment to create and ensure safe and healthy work by preventing all (serious) accidents, harm and work-related ill-health as well as promoting and achieving excellence in Safety, Health and Wellbeing at work (SHW).* It expresses the ambition that every accident or case of work-related ill-health is one too many. In this respect, it is a normative statement rather than a SMART target. Pointing out a clear direction (towards zero) is thereby essential. That requires a visible commitment of senior leaders and active participation and commitment of all internal and sometimes external stakeholders.

Vision is by definition future-oriented. Vision has been identified as a key element of transformational leadership for proactivity. Vision provides a motivational force for proactive action.[1] Vision Zero is an inspiring (long-term) goal and motivational force for many organizations and individuals. It implies that zero is a (long-term) aspirational goal.

Fifty years of goal-setting theory has resulted in abundant evidence that high or difficult goals generate the highest level of performance.[2] High and clear goals give direction: they direct attention and effort towards goal-relevant activities, energize people and facilitate persistence in case people are allowed to spend time to achieve their goals.

At the same time, Vision Zero is a transformational approach to doing business. SHW prevention is thereby regarded as an investment, not a cost factor. *Vision Zero is relevant for transforming business and employment by putting people at the centre of concerns and by greatly reducing work-related accidents and ill-health while making businesses more competitive and creating employers of choice.* Vision Zero is based

DOI: 10.1201/9781003621225-3

on the assumption that businesses do not only have financial goals but also contribute to a better and more sustainable world. This implies the integration of environmental, social and governance (ESG) aspects into business processes; for Vision Zero the focus is thereby on the social aspect. Vision Zero contributes to good and sustainable employership. Vision Zero also implies the responsibilities and commitment of the workforce. In this way, Vision Zero also contributes to good and sustainable employeeship.

Safety, health and wellbeing are more than rational goals; they are also human and social values and have an ethical or moral dimension. SHW belong to what most people judge to be important in life, for them personally as well as for their families and their colleagues. Vision Zero is an expression of the ethical notion that harm to people due to the work they do is not acceptable.[3]

> Values are important for each and every individual. Values provide guidance for people on what is the right thing to do and what not. Values underlie decisions and habits in organizations as well as societies. Values are closely associated with social norms and co-determine organizational cultures and subcultures.

The value of SHW is probably also why the ILO defined the right to a safe and healthy working environment as a fundamental right at work.[4] Like every human right, a safe and healthy working environment is based on human and social values. The same value aspects are associated with the older concept of 'decent work'[5] – a key element of the UN's Sustainable Development Goal 8: Decent work and economic growth.[6]

Besides the fact that the three aspects of safety, health and wellbeing at work are human and social values on their own, there are four (ethical) principles relevant for Vision Zero:[7]

- People have the fundamental right to a safe and healthy working environment
- Every worker should return home safely and healthy after work
- Human life is not negotiable; life and health are not for sale; the value of human life should not be expressed in monetary terms
- The physical and mental limits of human beings should be respected

Seeing SHW as a positive value is associated with attention to what can go wrong (accidents, incidents, cases of physical or mental ill-health, etc.). It implies that Vision Zero combines the promotion of positive approaches with (proactive) prevention of negative impacts. Adopting SHW as core values for the organization implies that these values play a role in decision-making processes and performing actions: they give *direction*. Values are a more fundamental basis for long-term policies than priorities, as priorities can easily change in case other important issues turn up. Similarly, a prevention culture based on these values is a better guarantee for continual improvement of the work environment than a priority setting based on ratio only. This is why in the European Union, it is an essential requirement for the offshore industry to have safety as a core value.[8]

A key characteristic of Vision Zero is the commitment to the *long-term goal*, which is clearly an ambition beyond compliance. Consequently, the change from *we have to* into *we want to* work safely and healthy is fundamental to Vision Zero. It is not only a matter of ratio and enlightened self-interest but also of internalizing SHW as human and social values. The internalization of 'we want to' is an expression of ownership and forms an essential aspect of a strong prevention culture.

Vision Zero builds on the knowledge and experience in the area of Occupational Safety and Health (OSH), including work engagement, workplace innovation, socio-technical design of organizations and high-reliable organizing (HRO). Safety, health and wellbeing at work are thereby regarded as both integrated and individual aspects. The three aspects of SHW have much in common, which forms the basis for synergy between these three aspects. Vision Zero also builds on similar initiatives in other areas, such as zero defects (total quality), net zero climate impact or zero waste and emissions (sustainability) – see also the section on the roots of Vision Zero (Section 3.2) and Section 10.1.

3.2 THE THREE ROOTS OF VISION ZERO

Vision Zero obviously builds on traditions and experiences with prevention in OSH. Apart from that, Vision Zero has three main roots: (1) the zero defects movement in total quality management – stemming from the world of business, (2) the development of a strong and learning safety or prevention culture – stemming from the investigation of disasters and (3) the (Swedish) policy and legislative initiatives for zero accidents in road safety. These roots come together in Vision Zero as we know it now.[9]

In the 1960s, the zero defects concept and approach were developed for quality management in the US defence and aerospace industries.[10] Leadership and a fruitful relationship between management and individual employees were regarded as the key to achieving zero defects. An essential principle is to *do the right things right from the start*, which implies serious attention to design, planning and preparation, including vocational training of people before they are assigned to a specific task. Zero Defects contributed to the success of the 'man on the moon' in 1969. It inspired industries in the early 1970s to start with zero accident campaigns, promoting the state of the art in accident prevention. Most of these programmes ended after a few years except in Japan where the zero accident campaign has been running since 1973. In the Japanese version, the participation of employees is seen as key to successful accident prevention.

In the aftermath of the nuclear disaster at Chernobyl (1986), a lot of attention has been paid to the importance of a strong safety culture for accident prevention.[11] The safety culture concept is useful for addressing intangible factors such as information difficulties, violations, failures to recognize emerging danger, role ambiguities, management complacency, poor communication and low prioritization of safety.[12] Values, alertness for upcoming or existing risks and mindsets are relevant intangibles. The safety culture concept was later broadened to the prevention culture concept to include all aspects of OSH, e.g., in the Seoul Declaration,[13] see Chapter 7.

FIGURE 3.1 The ISSA logos for Vision Zero, emphasizing its relevance for safety, health and wellbeing.

In 1997, the Swedish parliament adopted a zero accident policy and legislation for road traffic. The aim was that nobody should be killed or seriously injured due to traffic accidents. The prime responsibility was thereby attributed to the institutions involved in the design (and maintenance) of the road transport system, not to the individual drivers; the focus is strongly on the prevention of fatal and other serious accidents. In Sweden, the zero accident policy in road safety was very successful and was followed up with similar legal 'zero initiatives' with a focus on fire safety, patient safety, workplace safety and suicide prevention.[14]

In 2017, the International Social Security Association adopted their Vision Zero strategy addressing safety, health and wellbeing at work and launched a global Vision Zero campaign. In this approach, the three roots are more or less integrated. In this book, the scope of Vision Zero is the same as for the ISSA and aims at Zero (serious) work-related accidents and ill-health and excellence in SHW. The relationship between Vision Zero and SHW is expressed figuratively in the ISSA logos, see Figure 3.1.

The three roots of Vision Zero represent three points of view, leading to different accents in operationalization and implementation. The first approach inspired by zero defects emphasizes the business and economic value as well as the importance of leadership and participation. The second approach inspired by prevention culture focuses on the human and social values underlying Vision Zero, as well as on leadership and the role of people, their mental models and behaviour, collectively as well as individually. The third approach, inspired by the Swedish legislation on road safety, emphasizes the role of the national infrastructure, focuses on the prevention of *serious* accidents and the responsibility of system designers for the resulting performance. It is best to combine the various qualities of these three approaches in an integrated way.

3.3 THE FOUR KEY CHALLENGES FOR VISION ZERO

It is easy to say for organizations that they strive towards zero accidents and zero work-related ill-health. But it is a long-term challenge requiring leadership commitment and perseverance. This is due to some fundamental issues that complicate the realization of Vision Zero:

1. Zero risk does not exist.
2. Human beings and organizations make errors.
3. We live in a world of change; the future will be different and cannot be predicted.
4. Short-term solutions often hinder more far-reaching long-term solutions.

(1) Zero risk does not exist, but it makes good sense to eliminate unnecessary risks and to reduce the remaining risks following the ALARA (As Low As Reasonably Achievable) principle. What is reasonably achievable may depend to some extent on the context and can change over time through innovation or changes in cultural acceptance. (2) Human beings and organizations do make errors. Again, it makes good sense to reduce the likelihood of human errors for instance by good ergonomic design, or by taking care of the good physical and mental condition of the workers. An error-free organization or person does not exist, so it is important to reduce the impact of human or organizational errors before errors can lead to serious SHW issues. In high-risk industries, there is a great emphasis on designing processes *fail-safe*. The focus thereby is on technical processes. When people working in a team commit an error, team arrangements may ensure to respond adequately to those errors before they lead to an accident. (3) Every organization or human being is confronted with change, planned and not planned. The future is always different from the present. We can learn from past experience, but we cannot predict the future. By using forecasting techniques, scenario thinking and developing relevant skills, we can prepare for the future, but not control it. Therefore, you need flexibility and be prepared for new developments. There are techniques to learn from the future while it emerges.[15] In a world of change, control is difficult to realize, while too much control will hinder innovation and adaptation. (4) Short-term solutions often imply rather superficial solutions, whereby underlying problems are not, or not sufficiently addressed. Short-term solutions usually require only first-order learning processes and minor adaptations, while long-term solutions may require more drastic changes, other mental models and higher-order learning.

Some people are so impressed by one or several of the four key challenges that they believe that 'zero' is neither feasible nor realistic. They thereby neglect that the core challenge of SHW management is and will always remain precisely that: creating SHW *despite* risks, errors, change and the need for short-term solutions. To tackle these challenging issues requires leadership, proactivity, new mental models, learning and innovation, among other things. It is important to organize prevention right from the start of processes, i.e., already in the stages of decision-making and design. It also requires processes of long-term continual improvement, innovation and collaboration.[16]

3.4 THE BUSINESS PERSPECTIVE

It was already stated that Vision Zero is not only a transformational approach to SHW and prevention but also to doing business. SHW prevention is thereby regarded as an investment, not a cost factor. Vision Zero is relevant for transforming business and employment by putting people at the centre of concerns. By greatly reducing work-related accidents and ill-health, while making businesses more competitive, innovative and employers of choice.

3.4.1 The Strategic Value of Vision Zero

Each of the three aspects of SHW has a close relationship with working smoothly and productively, and the avoidance of undesirable losses. It should therefore be no surprise that SHW is becoming increasingly of strategic importance to organizations. However, the strategic importance of SHW is often not well understood. Relatively few organizations have defined strategies and long-term goals for these three aspects combined with their envisaged potential business benefits. There are four themes that highlight that strategic business value goes hand in hand with good SHW performance: reputation, people, culture and corporate social responsibility.

1. Reputation
   Good SHW contributes to an excellent reputation of the organization and its brands. An excellent reputation is conducive to being an attractive business partner and strengthens the positioning of the brands in a competitive market. An excellent reputation contributes to a good relationship with the authorities and a range of other stakeholders. That often results in greater understanding and more flexibility from their side. An excellent reputation in the labour market is conducive to becoming an employer of choice, which implies a greater attractiveness for talented workers. Finally, an excellent reputation of the organization often goes together with an excellent reputation of its senior managers, which is not only nice for their ego but also good for their income and their status in society.
2. People
   People are the most important asset of an organization – the only asset that cannot be copied by the competition. The strategic business value can be expressed in terms of increasing human, social and intellectual capital. To become an employer of choice means a lasting competitive advantage.
3. Culture
   A strong SHW culture is conducive to creativity, learning and innovation. That is not only relevant for improving SHW and dealing with unforeseen circumstances but also for business purposes and identifying and using business opportunities.
4. Corporate Social Responsibility (CSR)
   Taking active care of the employees, sharing values and creating a culture of prevention contribute to the social aspects of CSR. It forms an excellent basis for corporate citizenship and also for managing the social aspects of the value chain (see Section 10.3).

These four related themes clarify that good SHW performance forms a basis for lasting competitive advantage and implies opportunities to outperform competitors.

3.4.2 SHW as Business Opportunities

Some people regard safety, health and wellbeing dominantly as a cost factor. Other people see business processes dominantly as a risk factor that may affect SHW negatively. However, good SHW may positively impact business results, and business processes may positively influence SHW performance. The interactions between SHW and business processes can be mutually positive; the interactions potentially imply opportunities for creating win-win situations. See Figure 3.2.

It is important to acknowledge the opportunities for synergies between business processes and SHW. Synergies are very helpful to convince (top) managers to invest in SHW. Through synergies, SHW becomes a business value. That is essential for making SHW an integrated part of doing business; only then SHW can be a business need, i.e., more than 'nice to have'. It goes without saying that SHW will then be considered in decision-making about business processes.

The potential synergies between business processes and SHW imply that investing in SHW is not a *zero sum game*; it requires a mental model that is open to win-win situations. Investments may lead to achieving win-win situations and business as well as SHW achievements beyond compliance. It is useful to consider those interactions and synergies, and the contexts that influence them. For example, ageing of the workforce has implications for the management of business processes as well as for SHW performance.

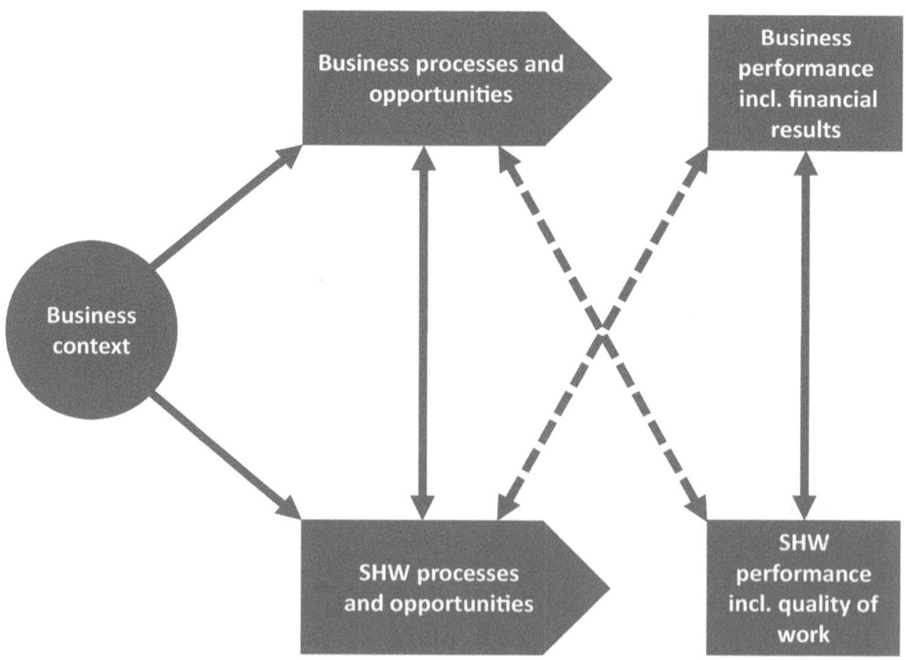

FIGURE 3.2 Synergies between business and SHW opportunities.

An interesting consequence of the interactions is that there are, in principle, two different groups of activities that can help to improve SHW: besides activities that are dedicated to SHW improvement in a direct way, activities focusing on business development may generate improvements in SHW as a side effect as well. A well-known example is the 'automation' of operational processes, which has eliminated repetitive and dangerous jobs in many production plants. Outsourcing dangerous activities to an organization specialized to do such work effectively and responsibly is another example. There are examples of negative interactions too: e.g., a business deal for additional work on top of existing activities may increase time pressure and may lead to significant overwork, both associated with several SHW risks. Note that due to the interactions between SHW activities and business processes, SHW can have a positive business impact, without realizing a significant positive impact on SHW. E.g., investments in SHW without much SHW effect, that nevertheless lead to greater attractiveness in the labour market or a better reputation.

3.4.3 Putting Vision Zero on the Business Agenda

Many organizations feel their identity and reputation do not allow for accidents, work-related physical or mental illnesses or unacceptable behaviours like discrimination or bullying. This implies the will to create organizations free of accidents and work-related illnesses and businesses where all employees can be confident that they will return home safely and in good health to their families.

Compliance is a legal and moral obligation associated with liabilities and the risk of fines and loss of reputation. It is therefore not surprising that approaches to SHW often focus on legal compliance. Though Vision Zero builds on the OSH tradition and knowledge, it differs from it. More specifically, Vision Zero builds on good practices in SHW, but its ambition goes beyond compliance. Therefore, it departs from a different mindset and requires a different type of actions and a long-term perspective. Paradoxically, for organizations going beyond compliance can be the shortest way to ensure compliance.

Safety is traditionally high on the agenda of the so-called high-risk sectors (industries with potential for disaster) such as the chemical or nuclear industry. This is because in these sectors, safety is regarded as vital for their 'license to operate', i.e., their reputation in society. However, in the knowledge sector, where services are the end product, and many jobs require brain power and creativity, (mental) health issues and the promotion of wellbeing are becoming increasingly important. This triggers knowledge businesses to pay attention to SHW and to go beyond SHW compliance. In a globalized world, businesses are confronted with many local and global competitors; investors increasingly apply ESG criteria; talented workers are increasingly mobile on the labour market; customers and end-consumers increasingly prefer products and services from organizations that take environmental and social sustainability seriously. This makes businesses increasingly aware of their corporate social responsibilities and the need to improve SHW.

Though the number of (serious) work-related accidents and diseases has certainly decreased during the past century, unfortunately, work-related accidents and diseases are still a tangible reality in many industries (see Chapter 1). In many organizations, there is still a world to win by implementing the good SHW practices that have been developed over the years. However, the actual dominant approaches to prevention are unable to achieve workplaces free of accidents and work-related ill-health: in many organizations committed to safety, the accident frequencies are plateauing or even going up; in many organizations physical and especially mental health issues are increasing, e.g., due to intensification of work.

There are many organizations that already have a strong focus on safety but have much less focus on health and wellbeing. Using the term 'health and safety' often means doing a lot for safety and only a little bit for physical or mental health. There are certainly organizations showing an excellent safety performance that still have quite a challenge to improve health and wellbeing. These organizations could use some of the strategies and programmes that contributed to safety excellence for similar activities to promote health and wellbeing. But the opposite is also true: strategies and programmes that are successful in promoting health and wellbeing may contribute to safety.

Modern organizations are now reflecting on the ethics that underpin their business. They are defining core values and developing strategies to flourish and survive in the long run from a business as well as a social community point of view. Vision Zero is then integrated into the technological transformations and organizational developments of the organization.

Vision Zero is not only for large multinational organizations, or for those organizations that already have an excellent SHW performance. In fact, any organization, including Small and Medium-sized Enterprises (SMEs), can start with a commitment to Vision Zero. It is not uncommon that organizations commit to Vision Zero after they have experienced a fatal or serious accident, or after several cases of serious work-related physical or mental diseases. Frequently organizations then suddenly realize the importance of managing SHW and want to make sure that serious accidents or work-related health problems will never occur again. See Section 9.6.

It goes without saying that adopting Vision Zero is only credible when it is associated with activities to improve SHW. Merely adopting and communicating Vision Zero as a slogan will not result in better SHW performance. When no improvements are experienced on the shopfloor, the Vision Zero initiative will, in the eyes of the workers, soon lose its credibility.

3.4.4 The Resource View on SHW and Good Business

In economic theory, the resource view says that the available resources determine the competitiveness of organizations. SHW is a factor that codetermines the functioning of people (human and social capital) and may contribute to the organization's principal value: a production factor to be developed and promoted.

As human, social and intellectual capital is increasingly important for many organizations; it is key to attract healthy and talented people. Nowadays, in many parts of the world, there is a shortage of professionals in areas that are key for competitiveness leading to the 'war for talent'.

The importance of SHW for positioning in the labour market can easily be overlooked or underestimated when the focus is limited to the SHW aspects.[17] Several relevant business benefits are *intangibles* such as becoming an *employer of choice* or improving *the employability of the workforce*. Unfortunately, these are difficult to value in financial terms and can be created only in the longer term. An adequate evaluation of such intangible benefits is therefore complicated. It is increasingly important to acknowledge that good SHW is a very important secondary working condition. Improving SHW implies that the organization becomes a more attractive employer, with positive effects on its capability to attract and retain professionals in their organization – a strategic business benefit.

At the same time, SHW has become a vital aspect of professionalism. Indeed, any good professional takes care of SHW at work. As Sennet says,[18] the craftsman is working professionally and knows the risks associated with the work and how to deal with them. During work, the professional is reflecting on his/her work implying that opportunities for (further) improvement are noticed, as well as potential new risks associated with work. This means that the craftsman/professional is able to continuously improve both SHW and the quality of work and the products or services that result from it.[19]

3.4.4.1 Safety as a Resource

Deviations in the production process imply undesirable costs and can trigger incidents. Good safety implies that production processes run smoothly, according to planning, without significant disruptions. This is why aiming at zero accidents is an integrated aspect of World Class Manufacturing[20] (see also Section 10.1).

There are increasing numbers of organizations that require their business partners to work safely – whether on the organization's site (labour contractors) or elsewhere (e.g., equipment suppliers). This implies that good safety has become an increasingly important business requirement; in this way, safety is a direct competitive factor.

Especially in high-hazard industries, improvements in safety may require substantial investments in hardware, while the safety benefits are a further reduction of the likelihood of a disaster which was already quite unlikely. In such cases, the investments in safety will usually not lead to short-term financial gains – more strategic benefits are important. The main benefits then are (1) exclusion (or significant reduction in likelihood) or severity of scenarios wherein business continuity is seriously threatened by a disastrous event, (2) a better relationship with governmental authorities (which in turn may lead to better opportunities for future business expansions) and (3) a better relationship with external stakeholders (e.g., neighbours). In case investment in hardware leads to more inherent safety (through measures 'at source'), it will reduce the ongoing costs for maintenance of additional (technological or organizational) measures. The Center for Chemical Process Safety (American Institute of Chemical Engineers) mentions five areas of benefit from Process Safety: corporate social responsibility, business flexibility, loss prevention, sustainable growth and leadership excellence.[21] For a comparison of the resource view with the risk view on safety, see Table 3.1.

TABLE 3.1
The resource versus the risk view on safety

The Resource View on Safety	The Risk View on Safety
Safety is a business resource: it contributes to a better reputation of organization and brand	Safety is a necessary condition for undisturbed production
Safety management focuses on the positive value of safety for business	Safety management focuses on risk reduction and problem-solving
Safety is associated with business benefits and clearly belongs to the core responsibility of managers	Safety is a side issue for businesses and regarded as of secondary importance
Safety contributes to opportunities for improvement and innovation	Safety management focuses on reducing the costs associated with accidents, incidents and near misses
Safety contributes to better relationship with a variety of stakeholders, including authorities	Safety is focused on compliance often associated with paper safety
A strong safety culture contributes to organizational learning and innovation	A calculating safety culture limits investments in safety
The business benefits of safety are acknowledged	Safety is dominantly seen as a cost factor
Costs for proactive and preventive safety are regarded as sound investments	Only strictly necessary costs are made for prevention, problem-solving and emergency preparedness

3.4.4.2 Health and Wellbeing as Resources

From a business perspective, health and wellbeing are closely associated with the ability to work, to be productive and be creative[22] – clearly the main business interests. Good (physical and mental) health is a precondition for being 'fit for work'. Good health is therefore necessary for optimum labour productivity. Illness leads to sickness, implying that workers cannot be productive and that other people (perhaps hired temporarily) have to take over their work. Wellbeing at work and good mental health are just as important for productivity. Especially in the modern economy, where a great part of the workforce is well-educated and brain work is more important than physical power, knowledge, thinking and creativity are key in many jobs. Good health and wellbeing are a prerequisite for the organization's ability to adapt and anticipate developments and external changes and a prerequisite for productivity and innovation.

A resilient and sustainable organization requires vital, flexible and creative people. While muscle power (the classic human resource) was essential in the early industrial society, we now have to acknowledge that brain power and intellectual capital are essential in the knowledge society, and social and communicative power and social capital are essential in a network society,[23] see Section 2.3. When health and wellbeing are regarded as resources instead of risks, this directly clarifies their importance for competitiveness. Organizations want to strengthen and develop their resources,

TABLE 3.2
The resource versus the risk view on health and wellbeing

The Resource View on Health and Wellbeing	The Risk View on Health and Wellbeing
Health is a personal and business resource for higher productivity	Health, i.e., being fit for work, is a necessary condition for productive employees
Health management focuses on the positive value of health and wellbeing for business	Health management focuses on risk reduction and on solving problems
Health is associated with organizational benefits and productivity that clearly belong to the core responsibility of managers	Health is associated with problems of individuals that managers do not like to feel responsible for
Health and wellbeing contribute to a healthy organization, also business-wise	Health management focuses on reducing the costs associated with unhealthy and less productive personnel
Health management implies taking care of health and vitality of all employees	Health management is taking care of people with health problems
The business benefits of good health and wellbeing are acknowledged	Health and wellbeing are dominantly seen as cost factors
Costs for proactive and preventive health activities are regarded as investments for a healthy organization	Only strictly necessary costs are made for prevention, cure, care, reduction of sickness absence and return to work

Source: Elaborated on Zwetsloot, G.I.J.M., Pot, F.D., (2004) The business value of health, *Journal of Business Ethics*, 55, 115–124. Doi: 10.1007/s10551-004-1895-9; Zwetsloot, G.I.J.M., (2019) Shared values for health, safety and well-being at work, in: Burke, R.J., Richardson, A.M., (eds), *Creating Psychologically Healthy Workplaces*. Cheltenham (UK): Edward Elgar. pp. 91–111.

and consequently employers have a clear interest in improving the health and wellbeing of their personnel: the promotion and development of the health and wellbeing of all personnel represents a business resource. Health problems can be a significant source of productivity losses,[24] and health risks imply potential business losses. For society at large, health and wellbeing are essential to keep sufficient numbers of people productive (at work) and for mitigating the skyrocketing costs of health care. It is interesting to compare the resource view with the risk view on health and wellbeing. See Table 3.2.

3.4.5 The Business Case/Benefits and Costs of SHW Improvement

Most managers have a natural interest in monetary benefits. Most senior managers are therefore fluently speaking the language of money, while most of them are less fluent in addressing the SHW aspects of doing business. At the same time, most SHW professionals are not that fluent in the language of business and money. This may hinder the development of genuine organizational commitment to improve SHW.

It is often stated that 'an ounce of prevention is worth a pound of cure'. Indeed, preventing a problem is often much cheaper than solving it. Insight into all costs and benefits may help to make rational decisions about investing in SHW. The costs associated with SHW are relatively easy to express in financial terms. The benefits associated with good SHW are, however, mostly intangible, difficult to quantify and difficult to express in financial terms. The combination thereof often results in an underestimation of the benefits and an overestimation of the costs of investing in SHW. This confirms the traditional mental model that SHW is associated with costs and that investments in preventive measures should primarily lead to cost reductions (e.g., less costs of sickness absence or financial losses due to disturbances in production). But in this way, the potential benefits of prevention in terms of added value are not included in the equation. When these benefits are seriously taken into account, it turns out that these are often quite significant.

Any accidental event, including minor incidents and near misses, implies more costs than most organizations are aware of. Even a minor incident leads to a disruption of the normal production process. Even when nobody is hurt and no material damage occurred, the work may be temporarily suspended, several people will be involved in discussing what has happened (i.e., not doing their normal work), trying to find out why and how it happened and how serious it could have been, and what could be done to prevent re-occurrence. All these costs are taken for granted and do not appear in the financial records. Compared to the costs of an incident, the costs for prevention are often insignificant.

It can be very useful to discuss the costs and benefits of a credible, potential worst-case scenario. For instance, when working with carcinogenic, reprotoxic or mutagenic substances. Consider a credible long-term health-related scenario, wherein potential health effects occur in members of the workforce, former personnel or personnel of contractors working for the organization. In case that scenario is realized, it will imply extremely high costs of legal affairs and compensation, with considerable loss of reputation due to social effects that are considered unacceptable in the society. Such a scenario will imply that effective preventive measures will have to be taken in any future similar risk-full situation, perhaps even in cases where serious impacts are less credible. Taken together, it will often be clear that taking the risk of not acknowledging a credible risk-full scenario implies an unacceptable business risk.

An important limitation of focussing on the costs associated with SHW is that these costs rely only partly on the employer. Depending on the jurisdiction, a significant part of the costs of material damage, individual injuries and damage to health is usually externalized, e.g., to the health care system, the social security system, insurers and the individual employees and their families. This may lead to decisions that look good from a narrow business perspective, but not good from the perspective of the workforce or society at large. This illustrates that SHW should be an integral part of the corporate social responsibilities of organizations. Fortunately, good SHW is often associated with business and societal benefits.

A side effect of investing in SHW is that people working in the organization will notice that the organization cares for them. Even apart from the real impact on SHW, the perception that the organization cares for them may in itself create better relationships between management and workers and improve the attractiveness of the employer in the labour market – an important business benefit.

In Table 3.3 examples are given of benefits in terms of added values created and reduced costs for businesses, SHW and society at large.

TABLE 3.3
Examples of added values and cost reductions generated through investment in SHW

Examples of Benefits of Investing in SHW			
Type of benefits	**Business/Economic benefits (examples)**	**Safety Health and Wellbeing benefits (examples)**	**Societal benefits (examples)**
Added values, i.e., benefits generated	Higher status in society; improved reputation of the organization and its brands preferred by important clients; higher stock market value; greater capability for innovation and change; creativity Higher reliability in the planning of work and projects Better relationship with authorities (license to operate) Improved reputation in the labour market offering an attractive working climate; becoming an employer of choice; more engaged personnel Higher labour productivity Higher product quality Greater organizational resilience	A vital and productive (ageing) workforce Being a safe and responsible employer and business partner High quality of work, e.g., opportunities for development and learning characterized by a high degree of self-organization Increased flexibility, employability and working ability of personnel	Decent employment and economic development A healthier and happier community, a decrease in health care cost More trust among social partners Higher percentage of productive people able to earn an income; less costs of early retirement due to loss of workability Increase in job security Increase in life expectancy
Type of costs	**Business/economic costs (examples)**	**SHW costs (examples)**	**Societal costs (examples)**
Cost reductions	Less legal costs and liability costs Fewer disruptions in the production process (productivity losses) Lower premiums for insurance Fewer costs of sickness absence and presenteeism Lower cost of temporary personnel replacing injured or sick people and for return-to-work programmes	Fewer incidents and accidents Fewer physical and mental health issues; fewer injuries Fewer costs of analysing incidents and other unplanned events Lower compensation costs (or premiums) Less medical costs	Lower costs of public health care and social security or workers' compensation Families affected (economically and socially) by injuries, diseases or illnesses of workers Medical care for employees

(continued)

TABLE 3.3 (Continued)
Examples of added values and cost reductions generated through investment in SHW

Type of costs	Business/economic costs (examples)	SHW costs (examples)	Societal costs (examples)
	Lower turnover of personnel, less costs of attracting and selecting new personnel Fewer costs for recruitment and retention of personnel; fewer costs for hazard pay Less opportunity costs (missed productivity, synergy, etc.)		

Source: Elaborated on Zwetsloot, G.I.J.M., van Scheppingen, A.R., (2007) Towards a strategic business case for health management, in: Johansson, U., Ahonen, G., Roslander, R., (eds), *Work Health and Management Control.* Stockholm: Thomson Fakta. pp 183–213.

While business cases usually have a short-term perspective (return on investment within a few years), value cases have a broader and long-term perspective. Value cases can be characterized by a combination of five perspectives: Costs and benefits for all relevant stakeholders, a broad value concept which includes also qualitative elements, the combination of a short- and long-term perspective, holistic thinking (systems perspective) and the inclusion of complex factors like interdependencies and uncertainties.[25]

Investing in the design stage, e.g., in ergonomic design of technical installations or workplaces is often financially preferable, as it generates benefits for the lifetime of the designed object or process. Due to a short-term perspective of a business case, such structural long-term benefits may remain neglected or undervalued. Economic incentives (e.g., tax reductions, subsidies) can make investments in SHW improvement financially more attractive; see Section 10.2.

3.5 GOAL ZERO

For Vision Zero, the conviction that all (serious) accidents and work-related ill-health can be prevented is essential. In hindsight, after an incident has occurred, it is relatively easy to suggest how it could have been prevented. But often it is a tough challenge to prevent incidents up-front.

Even when SHW is regarded as a core value, it is clear that, in order for an organization to survive Vision Zero cannot have always an absolute priority. Vision Zero is not absolute, as that would most probably imply an immediate stop of production activities as these always imply certain risks. But for Vision Zero, trade-offs are not made in the stage of formulating ambitions and goal setting, but only considered in the phases thereafter, i.e., during operationalization and implementation However,

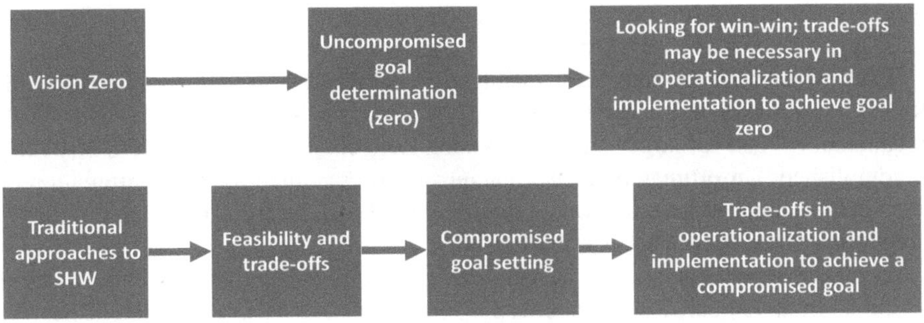

FIGURE 3.3 Trade-offs in goal setting and implementation in traditional SHW and Vision Zero. (Inspired by Abebe, H.G., Hansson, S.O., Björnberg, K.E., (2022) Arguments against Vision Zero: A literature review, in: Edvardsson Björnberg, K., Belin, M.Å., Hansson, S.O., Tingvall, C., (eds), *The Vision Zero Handbook*. Cham: Springer. Doi: 10.1007/978-3-030-23176-7_3-1

the mindset is to focus on win-win situations. Only when that is not a realistic option, trade-off may be necessary during the operationalization of the zero goal and the implementation of the Vision Zero strategy. See Figure 3.3.

In traditional SHW management, trade-offs between production and safety and health are found in all stages: in goal setting, operationalization of the goals and implementation. As a businessman said regarding negotiations for a business deal: 'if we dilute wine twice with water, the result will be water with only a slight taste of wine'.

Vision Zero is not the only challenging Vision with a 'zero-type goal': the United Nations Sustainable Development Goal (UN SDG) #1 reads 'no poverty' and UN SDG #2 reads 'zero hunger'. The fundamental principles and rights at work as defined by the ILO comprise the 'abolition of child labour' and 'elimination of discrimination in respect of employment and occupation'. Several parts of the Universal Declaration on Human Rights address zero-type goals, for example: 'No one shall be held in slavery or servitude'. The concept of World Class Manufacturing (WCM) comprises 'zero defects', 'zero delays', 'zero accidents' and 'zero waste'.[26] There are also examples of zero in everyday life: Cola-zero (zero sugar), zero tolerance against unacceptable behaviour, zero kilometres (for locally produced food), zero-emission cars, paperless offices, contactless payments, etc.

Paradoxically, the most important challenge for Vision Zero is *not* to reach 'zero'. Accident frequencies and work-related ill-health are outcomes of production and work processes. But (accidental) outcomes are not really manageable; instead, production and work processes that generate these outcomes can and should be managed. Counting accidents and work-related ill-health is looking at past performance and has little predictive value. The Vision Zero concept is future-oriented and therefore refers to manageable actions, which enhance the controllability of operations. The strength of Vision Zero lies partly in the fact that many organizations and people have experienced accidents and work-related ill-health by themselves or others: they want to get rid of them. Zero accidents and work-related ill-health give a more concrete picture of good SHW than the abstract positive concepts of safety, health and wellbeing themselves. Moreover, loss aversion implies that 'The response to losses is stronger than the response to corresponding gains'.[27]

Even when organizations have managed to achieve zero incidents and ill-health for a certain period of time (which for statistical reasons is easier to achieve for small organizations than for larger ones), the Vision Zero challenge is not necessarily fully accomplished. Unfortunately, achieving zero is no guarantee for the continuation of such achievement. The reduction of the likelihood and severity of possible undesirable events remains an ongoing challenge.

In reality, there is no sound alternative to goal zero. What could be a credible and useful alternative goal? Five accidents per year? Ten cases of work-related ill-health? You cannot make a SMART plan anyway to realize a desired number of accidents or ill-health. When they occur, it is despite the plans made. Plans for reducing the number of accidents to X are never SMART. You can only make a SMART plan for controllable preventive action. It is therefore much better to measure performance by proactive leading indicators (see Section 9.2) than to refer to lagging indicators like the number of accidents or the frequency of Lost Time Injuries (LTIs).

The *zero* in Vision Zero represents an aspirational goal, not a target. Zero as a target would have several disadvantages. (1) Lagging indicators reflect past performance, not the actual or future performance. (2) There is always certain statistical noise in accident and health data, which implies that the results will inevitably fluctuate. (3) When a target is suddenly not met in a certain department, the statistical process known as 'regression to the mean' is likely to lead automatically to better figures, without actually realizing improvements in the processes that generate them. (4) Taking measures after each undesirable event may disturb the system and can be counter-productive. Finally, (5) in safety literature it is emphasized that reporting accidents can easily be discouraged to get nicer figures. Each of these disadvantages is not unique to 'zero' as a target, in fact, they are relevant for *any* ambitious quantitative target. Deming[28] already noted these kinds of negative impacts and saw them as inherent in Management by Objectives (MBO); Deming regarded MBO as a serious business disease that inhibits processes of continual improvement.

An interesting discussion is whether the ambition and commitment of Vision Zero should be to prevent *all* work-related accidents and ill-health or to prevent *all serious* work-related accidents and ill-health. A lot of attention to minor issues may imply that too many resources (time, money, effort) are spent to deal with them, at the expense of attention to more serious risks. From an ethical or value point of view, the ambition should clearly be to prevent all accidents and ill-health, but at the same time it is crystal clear that serious issues are more important than minor issues. The quality guru Joseph Juran formulated the solution to this dilemma: it is important to distinguish *the vital few* from *the trivial many*. Each serious risk belongs to *the vital few* and deserves dedicated prevention measures, leadership involvement, training, programmes, etc., minor issues belong to 'the trivial many'. The minor issues that form *the trivial many* should be approached as a group. The group of the trivial many should be treated as of equal importance as a serious issue. For SHW, this may imply that attention to the multitude of minor SHW risks is integrated in an SHW management system and the promotion of a strong prevention culture.

VISION ZERO IN ROAD SAFETY: A CLEAR FOCUS ON THE PREVENTION OF SERIOUS ACCIDENTS

In the area of road safety, where Vision Zero is adopted in many countries, regions and cities, the choice has been clearly made to focus on the prevention of fatalities and other serious accidents and to see reduction of minor accidents as a side issue. Consequently, the focus is on reducing the severity of accidents (e.g., by making a seat belt mandatory) or on reducing the likelihood of severe accidents (e.g., by reducing the alcohol consumption of drivers or discouraging the use of smartphones while driving). The latter type of measures reduces the likelihood of minor accidents as well. Interesting is the proliferation of roundabouts at dangerous crossings. They significantly reduce the number of fatalities and severe accidents, while *increasing* minor accidents. From a Vision Zero road safety point of view, roundabouts are very successful.

Vision Zero does not mean that disturbances or unexpected or undesirable events cannot take place. But it implies that such events are anticipated as early as possible and responded to as soon as possible. This then results in proactive prevention or actions for improvement.

3.6 INNOVATING TO ZERO

For today's businesses, there are several 'Zeros' to aim at. Many leading organizations have committed themselves to zero defects, zero emissions, zero traffic accidents, zero environmental wastes or zero economic wastes (lean). Each of these visions was initially criticized for being unrealistic, too expensive and not feasible. But each of them triggered many successful innovations associated with *good business cases*.[29] Most of these *aims to zero* are supported by governments, NGOs or (international) agreements between organizations working in certain industries. The European Union strives for zero work-related deaths.[30]

It is important to be aware that simply to keep on doing the same things better than before will not be a successful strategy to achieve zero; see Section 8.1.2. Trying harder is not likely to work either; it would probably lead to greater resistance as well as to a decrease in credibility and efficiency of efforts made. Instead, innovation and change are essential, as well as out-of-the-box thinking and higher-order learning for solving existing problems in innovative ways. Business strategist Singh identified *innovating to zero* as a mega trend that is influencing business, work and life in many areas.[31] Singh regards *innovating to zero* as a journey which implies huge challenges and opportunities and requires an innovation agenda that focuses on breakthroughs in the face of radical goals. This also applies to Vision Zero.

3.6.1 The Seven Innovative Qualities of Vision Zero

When we take a closer look at the concept of Vision Zero, seven innovative qualities of Vision Zero come to the forefront. These qualities have to be developed while implementing Vision Zero. They can be understood as opportunities for the process of continual improvement and innovation towards the realization of Vision Zero.

1. *Safety, Health and Wellbeing are approached in an integrated way.* That transcends the limits of the silos of the respective disciplines; instead, it will facilitate the search for synergies between the three aspects. The integrated approach appeals usually more to higher managers and policymakers who have an integrated responsibility than to the respective experts.
2. *Vision Zero requires new mental models.* The ambition to prevent all (serious) accidents and work-related ill-health is a bold one. Therefore, Vision Zero implies a long-term ambition that clearly goes beyond the present state of the art. It is known that when you continue to do the things you already did, you will achieve similar results, not significant improvements. The same is true for our thinking. When we continue to think about SHW in the way we have done for decades, we cannot expect to realize much progress, see Section 4.6.
3. *Vision Zero has a notable value dimension.* Respecting and protecting human life and health at work is essential for Vision Zero. SHW represents essential human and social values. The adoption of these values (and some supporting values) is important for being a good employer as well as for being a good employee.
4. *Vision Zero implies a long-term perspective.* The prevention of all (serious) accidents and work-related diseases will require long-term processes, requiring strategy and perseverance. Short-term solutions often impede or discourage investments in long-term solutions. The long-term perspective therefore requires to reconsider traditional approaches and to re-evaluate interventions by no longer considering only their short-term effects but their long-term impacts too.
5. *Vision Zero implies going beyond compliance.* Consequently, its implementation is not a matter of using power and coercion. It requires the change from *we have to* into *we want to.* This is essential for proactivity, creativity, ownership and intrinsic motivation. These cannot be forced, only enabled and promoted.
6. *Vision Zero requires commitment, collaboration and collective learning.* At an organizational level, commitment, cooperation and organizational learning apply to the majority of managers and workers, but it may include external stakeholders like suppliers and contractors. At the country or international level, it requires the participation, collaboration and collective learning of a variety of stakeholders.[32]
7. *Vision Zero aims at the prevention of accidents and work-related ill-health as early as possible.* This requires consideration of SHW aspects in the earliest possible phase; i.e., right from the start of any process. It implies an orientation on the future and using foresight techniques. The focus should not be

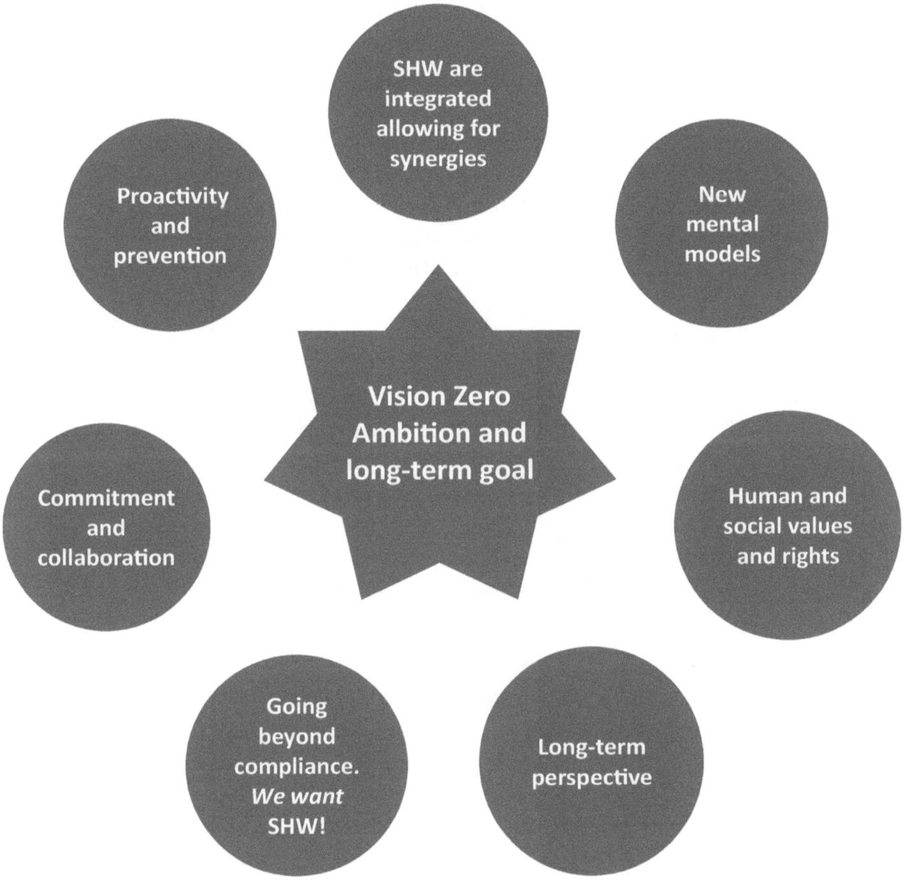

FIGURE 3.4 The innovative qualities of Vision Zero.

limited to the stage of production and work; it is necessary to pay structural attention to the earlier stages of decision-making, design and development.

Each of the seven innovative potentials mentioned above is essential for Vision Zero. Jointly, they make Vision Zero interesting for organizations, policymakers, consultants and researchers, see Figure 3.4.

3.7 THE COMMITMENT STRATEGY

Many authors have identified management commitment as a crucial factor for effective SHW strategies.[33] Because Vision Zero requires a long-term improvement process, committed leadership at the top of the organization is a vital condition. Otherwise, perseverance is impossible. In practice, commitment to safety creates better risk reduction and control strategies, not the other way around.[34]

Leadership commitment to Vision Zero is key to give and maintain direction in SHW and to stimulate and support proactive prevention measures (see Chapter 5). The commitment of the leaders is based on SHW as values and the importance of business continuity. The commitment of the workforce is based not only on well-understood self-interest in good SHW but also on the perception of a certain degree of autonomy, social support by leaders and colleagues and competence. These three latter factors represent basic human needs and, according to the self-determination theory,[35] they are essential for supporting intrinsic motivation and work engagement.[36] Proactive behaviour requires intrinsic motivation and cannot be triggered by coercion or economic incentives. Work engagement has positive effects on wellbeing as well as on safety and health and is important for the commitment strategy.

Research on the so-called **goal setting theory** has clarified that the commitment to a goal can be enhanced by leaders who communicate an inspiring vision and behave supportively.[37] When people are committed to goals, this triggers higher performance. Goal commitment is facilitated by making the goal important to people and the belief that they can attain the goal; that is enhanced by adequate training, role modelling and persuasive communication.

Vision Zero puts people at the centre of concerns and underlines practicing social and human values as business values. The commitment strategy is effective both in stable and dynamic complex situations, especially when the ambitions are 'going beyond compliance'. The strategy builds on values as well as ratios and is useful for dealing with complexity and variability. It is based more on confidence, cooperation and learning than on procedures and compliance. This implies that commitment strategies are characterized by promoting and strengthening ownership as well as individual and collective learning from errors and events like accidents and incidents. The commitment strategy is known to be successful in generating high (business) performance.[38] Commitment is especially relevant to support proactivity, creativity, (higher-order) learning and innovation; it also stimulates bottom-up initiatives.

This is why for organizations dealing with complex processes, the commitment of every manager and worker is nowadays regarded as an essential feature for successful and resilient organizations.[39] Vision Zero provides a clear message from the top leaders of the organization, and it can boost the development of a broad prevention culture. As a logical consequence, Vision Zero requires the commitment of middle- and lower-level managers and the workforce as well. Commitment to SHW cannot be generated by rational argumentation alone: it is also a matter of business ethics.

A high-commitment high-performance strategy is characterized by a positive reciprocal relationship between the goals and purpose of an organization and the goals and purposes of its workforce. This implies the commitment of top management to the needs and safety, health and wellbeing of the workers and the commitment of the workers to the goals of the organization (see Section 5.5). Workers are then internally motivated to achieve the organizational goals and *own* them. This is a crucial prerequisite for Vision Zero to succeed. Without commitment and ownership, it is impossible to make the essential cultural change from *we have to* into *we want to*. That is supported by shared values between (top) management and workers. This requires mutual trust, psychological safety and the absence of fear. Sharing values and purpose strongly reduces the need for control. It will contribute to psychological safety, i.e., the absence of fear and anxiety which is important for open and honest communication and mutual learning, e.g., on unexpected or personal problems encountered and reporting of accidents. A commitment strategy is always associated with a high degree of social capital and a strong community.

Research on the implementation of the Zero Accident Vision (ZAV) showed that it led managers to give safety a very high priority in daily practice, associated with higher worker commitment and more empowerment.[40] That goes together with encouragement of participation and learning and a culture that has many characteristics of a 'just culture'.[41]

When there is a genuine commitment to SHW, suggestions or opportunities for improvement will be taken seriously. People will feel encouraged to come up with suggestions for improvement or to ask questions in case of doubt. They are confident to receive timely and serious feedback from their leaders. People will feel free to report near misses, errors or unexpected phenomena and to discuss SHW in case of doubt.

Commitment strategies have been applied in human resource management since the mid-1980s[42] and are characterized by the combination of high commitment and high performance. Commitment and engagement allow for a minimum of administrative or hierarchical controls. Interesting is the finding of Edwards and Jabs[43] that a safety culture that promotes greater commitment and empowerment is not compatible with traditional bureaucratic controls. That has its consequences for the implementation of Vision Zero.

Today, a risk control strategy for managing SHW is the dominant approach in most organizations. This strategy focuses on rational aspects of production and compliance with regulations, standards and the like, in order to keep a license to operate. This approach starts with the identification of hazards and risks and is followed by planning and implementation of measures to control those risks that are regarded as unacceptable. The focus is often on solving existing SHW problems. This implies that when the most pressing problems (those severe or urgent) are solved, the remaining problems will receive less priority. In this way, the risk management approach is often self-restricted by limited priorities for further improvements. A risk control strategy is therefore not successful in realizing long-term continual improvements in SHW. Up to the level of compliance, the risk control strategy is usually effective in situations that are stable and not very complex, but a risk control strategy is not effective in going beyond compliance. Such a strategy cannot be successful in case of high levels of uncertainty, ambiguity or complexity. Instead, dynamic risk control in complex situations is compatible with a commitment strategy.

Vision Zero builds on SHW as values as well as enlightened self-interest of the organization and its members. Personal commitment of (top) managers is the basis for genuine visible leadership. The majority of the workforce feels ownership which is associated with intrinsic motivation to contribute to SHW. In this way, commitment forms the basis for proactive behaviour, creativity, learning, innovation and the development of a strong prevention culture. The high level of commitment creates a culture wherein *we want excellent SHW* is the leading motivation, whereas in more traditional cultures the focus is on SHW as a set of externally defined obligations.

3.8 THE ZERO DEBATE

Leaders of organizations are used to think in terms of strategy and long-term goals. They understand the value of steering on ambitions and accept that there are always uncertainties associated with it. For them, Vision Zero is a logical and attractive concept. Therefore, higher managers are usually enthusiastic about the ambition of Vision Zero. But other people may show resistance to Vision Zero for a variety of reasons, mostly based on a misunderstanding of Vision Zero, for example, the idea that Vision Zero implies zero risks and absolute safety or that Vision Zero does not allow for trade-offs (compare with Section 3.5). Some people criticize Vision Zero because they assume it is a target (instead of an ambition). Others assume that Vision Zero implies more rules and behavioural control which would make it paternalistic and would restrict the freedom of behaviour of the workforce. But in order to successfully implement Vision Zero, it is vital that an organization develops a sound commitment strategy, wherein all stakeholders are engaged and participate actively and proactively. That requires a process to share values and develop a more proactive mindset, certainly not a paternalistic process. In fact, introducing Vision Zero builds on the interests of all members of an organization. As soon as the long-term benefits are understood and some short-term actions show it is to be taken seriously, initial resistance will change into enthusiastic commitment.

3.8.1 Critical Arguments That Helped to Strengthen Vision Zero

Just like in science, it is important that sound critical arguments should not be ignored. Of course, sound criticisms on Vision Zero are possible and may point out weaknesses in the Vision Zero strategy. By acknowledging such weaknesses, the approach was already adapted and made more robust.

There are three critical statements to consider seriously in this respect: (1) Zero work-related accidents or illnesses is not a good measure for SHW. (2) The pressure to achieve 'zero' can make people hesitating to report incidents. Manipulation of SHW data is sometimes easier than managing prevention. (3) Preventing all incidents requires too much effort for minor SHW risks, thereby wasting resources. The critical issues have already contributed to the further development of Vision Zero.

1. Given the human toll of work as described in Chapter 1 and the reality that work-related accidents and diseases are rather common, zero is clearly an indication of the 'right direction'. But striving for *zero* is indeed somewhat paradoxical. Zero work-related accidents or illnesses are not a good measure for SHW. Reaching 'zero' is not a good indicator for SHW. In Section 3.5

this was already addressed. It is one of the reasons to promote the proactive leading indicators, see Section 9.2.

2. The pressure to achieve 'zero' can make people hesitating to report incidents. Manipulation of SHW data is sometimes easier than managing prevention. That may generate scepticism in the organization and undermine the credibility of Vision Zero. It is therefore important *not to make zero a target* and not to make people or business units accountable for achieving zero. It is certainly possible to make people accountable in a Vision Zero strategy but that should refer to their preventive efforts, for instance, by using proactive leading indicators as we will see in Section 9.2.
3. Prevention of all work-related accidents and ill-health incidents implies that a lot of effort is needed for the prevention of minor issues. That may imply a waste of effort, spending too much resources for relatively minor achievements. As discussed in Section 3.5, minor issues should be treated groupwise; that group does deserve attention similar to each of the serious risks. The seven Golden Rules which will be discussed in Section 9.1 are extremely useful in this respect.

3.9 CONCLUSIONS

Vision Zero is the ambition and commitment to create and ensure safe and healthy work by preventing all (serious) accidents, harm and work-related ill-health, and promoting excellence in Safety, Health and Wellbeing at work (SHW). It departs from the recognition of a safe and healthy working environment as a fundamental labour right. Vision Zero is also a transformational approach to doing business wherein people are put at the centre of concerns, and SHW is regarded as an investment, not a cost factor. Organizations feel their identity and reputation do not allow for accidents, work-related ill-health or unacceptable behaviours like discrimination or bullying. Each of the three aspects – safety, health and wellbeing – has potentially a positive business value. Strategic business benefits of Vision Zero stem from the impact on reputation, people, an innovative prevention culture and corporate social responsibility.

Vision Zero is increasingly regarded as the approach to realize long-term and lasting significant improvements in SHW. SHW are regarded as (positive) human and social values to be promoted but also as areas requiring structural prevention. Vision Zero is a bold and long-term ambition which has to overcome four major challenges: risks, errors, change and the propensity to prefer short-term solutions. Such ambitions cannot be achieved by doing the same things as before, or by thinking in the same way as before. A crucial difference with more traditional ways of goal setting and implementation is that trade-offs are not made in the stage of formulating ambitions and goal setting. The zero in Vision Zero is an aspirational goal, not a target. Paradoxically, for Vision Zero the most important challenge is not to reach 'zero'. For Vision Zero the principal performance measures are proactive leading indicators.

Genuine commitment from top management to the workforce implies ownership of improvement and learning processes. That forms the basis for a commitment strategy and an 'innovation to zero' process. Vision Zero requires genuinely committed leadership, consequent prevention and the development of a strong and learning prevention culture. It implies a change from compliance – *we have to* – external motivation, into commitment and participation – *we want to* – intrinsic motivation. This implies a culture of ownership, proactive behaviour, learning, creativity and innovation.

Seven innovative qualities of Vision Zero imply opportunities for innovation and should be explored: (1) Safety, health and wellbeing are approached in an integrated way, allowing synergies; (2) Vision Zero departs from new mental models; (3) Vision Zero has a notable value and rights dimension; (4) It implies a long-term perspective; (5) It implies going beyond compliance; (6) It is based on commitment and collaboration; and finally (7) It is proactive and aims at prevention as early as possible, i.e., already in the stages before production starts.

In the next chapter, a number of additional concepts and theories that underlie Vision Zero will be presented and discussed.

NOTES

1 Griffin, M.A., Parker, S.K., Mason, C.M., (2010) Leader vision and the development of adaptive and proactive performance: A longitudinal study. *Journal of Applied Psychology*, 95 (1), 174–182. Doi:10.1037/a0017263.

2 Goal-setting theory. See Locke, E.A, Latham, G.P., (2022) Building a practically useful theory of goal setting and task motivation: A 35-year odyssey. *American Psychologist*, 57 (9), 705–717. Doi:10.1037/0003-066X.57.9.705.

3 ILO, (2022) *ILO Declaration on Fundamental Principles and Rights at Work and Its Follow-up* (amended version). Geneva: International Labour Organization; Zwetsloot, G.I.J.M., van Scheppingen, A.R., Bos, E.H., Dijkman, A., Starren, A., (2013) The core values that support health, safety and well-being at work. *Safety and Health at Work,* 4, 187–196. Doi: 10.1016/j.shaw/2013.10.001.

4 ILO, (2022) *ILO Declaration on Fundamental Principles and Rights at Work and Its Follow-up* (amended version). Geneva: International Labour Organization.

5 Ghai, D., (2008) Decent work, concept and indicators. *International Labour Review*, 142, 113–145.

6 UN, (2015) *United Nations Sustainable Development Goals*.

7 Eichendorf, W., Bollmann, U., (2014) Future approaches to a culture of prevention. In: *From Risk to Vision Zero – Selected Proceedings of the International Symposium on Culture of Prevention – Future Strategies*, 43–53, Helsinki: FIOH.

8 EC, (2013) *European Directive on Safety of Offshore Oil and Gas Operations*. Directive 2013/30 of the European parliament and the council, article 8, clause 6. Brussels: European Commission.

9 Zwetsloot, G.I.J.M., Kines, P., (2022) Vision Zero in workplaces. In: Edvardsson, Björnberg, K., Belin, M.Å., Hansson, S.O., Tingvall, C., (eds) *The Vision Zero Handbook*. Cham: Springer. Doi: 10.1007/978-3-030-23176-7_41-1.

10 Halpin, J.F., (1966) *Zero Defects – A New Dimension in Quality Assurance*. New York: Mc Graw-Hill.

11 INSAG, (1986) *Summary Report on the Post-Accident Review on the Chernobyl Accident*. Safety Series No. 75, International Nuclear Safety Advisory Group. Vienna: IAEA.

12 Cox, S.J., Flin, R., (1998) Safety culture: philosopher's stone or man of straw? *Work and Stress*, 12 (3), 189–201. Doi: 10.1080/02678379808256861.

13 ILO, (2008) Seoul Declaration on Safety and Health at Work. Seoul (29 June 2008): ILO, ISSA and KOSHA.

14 Kristianssen, A.C., Andersson, R., Belin, M.A., Nilsen, P., (2018) Swedish Vision Zero policies for safety – a comparative policy content analysis, *Safety Science*, 103, 260–269. Doi: 10.1016/j.ssci.2017.11.005.

15 Scharmer, C.O., (2007) *Theory U – Leading from the Future as It Emerges*. Cambridge, MA: SoL.
16 Alanko, T., Ruotsala, R., (2022) A multi-perspective framework of Vision Zero: Toward collaborative promotion of safety, health and wellbeing at work. *Safety and Health at Work*, 13, 372–375. Doi: 10.1016/j.shaw.2022.05.001.
17 Zwetsloot, G.I.J.M., van Scheppingen, A.R., Dijkman, A., Heinrich, J., den Besten, H., (2010) The organizational benefits of investing in workplace health. *International Journal of Workplace Health Management*, 3 (2), 143–159. Doi: 10.1108/17538351011055032.
18 Sennet, R., (2008) *The Craftsman*. New Haven: Yale University Press.
19 Sennet, R., (2008) *The Craftsman*. New Haven: Yale University Press.
20 D'Orazio, L., Messina, R., Schirald, M.M., (2020) Industry 4.0 and world class manufacturing integration: 100 technologies for a WCM-I4.0 Matrix. *Applied Sciences*, 10, 4942. Doi: 10.3390/app10144942.
21 CCPS, (2018) *The Business Case for Process Safety*. 4th edition. New York: Center for Chemical Process Safety.
22 Zwetsloot, G.I.J.M., Pot, F.D., (2004) The business value of health, *Journal of Business Ethics*, 55, 115–124. Doi: 10.1007/s10551-004-1895-9.
23 Zwetsloot, G.I.J.M., van Scheppingen, A.R., (2007) Towards a strategic business case for health management, in: Johansson, U., Ahonen, G., Roslander, R., (eds), *Work Health and Management Control*. Stockholm: Thomson Fakta. pp. 183–213.
24 Zwetsloot, G.I.J.M., Pot, F.D., (2004) The business value of health, *Journal of Business Ethics*, 55, 115–124. Doi: 10.1007/s10551-004-1895-9.
25 Van Scheppingen, A.R., Baken, N., Zwetsloot, G.I.J.M., Bos, E.H., Berkers, F., (2012) A value case methodology to enable a transition towards generative health management: A case study from the Netherlands, *Journal of Human Resource Costing and Accounting*, 16 (4), 302–319; Jain, A., Leka, S., Zwetsloot, G. I. J.M., (2018) The economic, business and value case for health, safety and well-being, in: *Managing Health, Safety and Well-Being. Dordrecht:* Springer. pp. 183–213. Doi: 10.1007/978-94-024-1261-1#toc.
26 D'Orazio, L., Messina, R., Schirald, M.M., (2020) Industry 4.0 and world class manufacturing integration: 100 technologies for a WCM-I4.0 matrix. *Applied Sciences*, 10, 4942. Doi: 10.3390/app10144942.
27 Kahneman, D., Tversky, A., (1979) Prospect theory: An analysis of decision under risk, *Econometrica*, 47 (4), 263–291. Doi:10.2307/1914185.
28 Deming, E.W., (1990) *Out of the Crisis*. Cambridge: Cambridge University Press (first edition, 1982).
29 Zwetsloot, G.I.J.M., Aaltonen, M., Wybo, J.L., Saari, J., Kines, P., Op De Beeck, R., (2013) The case for research into the zero accident vision, *Safety Science*, 58, 41–48. Doi: 10.1016/j.ssci.2013.01.026; Jain, A., Leka, S., Zwetsloot, G. I. J.M., (2018) The economic, business and value case for health, safety and well-being, in: *Managing Health, Safety and Well-Being*. Dordrecht: Springer. pp. 67–98.
30 EC, (2021) European Commission. *EU Strategic Framework on Health and Safety at Work 2021-2027 – Occupational Safety and Health in a Changing World of Work*. Brussels, European Commission, COM (2021) 323 final.
31 Singh, S., (2012). *New Mega Trends – Implications for Our Future Lives*, Hampshire (UK), New York (US): Palgrave Macmillan.
32 Alanko, T., Ruotsala, R., (2022) A multi-perspective framework of Vision Zero: Toward collaborative promotion of safety, health and wellbeing at work. *Safety and Health at Work*, 13, 372–375. Doi: 10.1016/j.shaw.2022.05.001.
33 Barling, J., Loughlin, C., Kelloway, E.K., (2002) Development and test of a model linking safety-specific transformational leadership and occupational safety. *Journal of Applied Psychology*, 87 (3), 488–496. Doi: 10.1037/0021-9010.87.3.488. See also Section 9.1.
34 Locke, R., Amengual, M., Mangla, A., (2009) Virtue out of necessity? Compliance, commitment, and the improvement of labor conditions in global supply chains. *Politics & Society*, 37 (3), 319–351. Doi: 10.1177/0032329209338922.
35 Ryan, R.M., Deci, E.L., (2000) The self-determination theory and the facilitation of intrinsic motivation, social development and well-being, *American Psychologist*, 55 (1), 68–78.
36 Van Tuin, L., Schaufeli, W.B., Van den Broeck, A., (2021) Engaging leadership: Enhancing work engagement through intrinsic values and needs satisfaction. *Human Resource Development Quarterly*, 32, 483–505. Doi: 10.1002/hrdq.21430.

37 Locke, E.A., Latham, G.P., (2002) Building a practically useful theory of goal setting and task motivation: A 35-year odyssey. *American Psychologist*, 57 (9), 705–717. Doi: 10.1037/0003-066X.57.9.705.
38 Beer, M., (2009) *High Commitment, High Performance – How to Build a Resilient Organization for Sustained Advantage.* San Francisco: Jossey-Bass.
39 Beer, M., (2009) *High Commitment, High Performance – How to Build a Resilient Organization for Sustained Advantage.* San Francisco: Jossey-Bass.
40 Zwetsloot, G.I.J.M., Kines, P., Ruotsala, R., Drupsteen, L., Merivirta, M.L., Bezemer, R.A., (2017) The importance of commitment, communication, culture and learning for the implementation of the zero accident vision in 27 companies in Europe, *Safety Science*, 96, 22–32. Doi: 10.1016/j.ssci.2017.03.001.
41 Zwetsloot, G.I.J.M., Kines, P., Ruotsala, R., Drupsteen, L., Bezemer, R., (2015) *Success Factors for the Implementation of a Zero Accident Vision (ZAV),* Report to the DGUV, Netherlands Organisation of Applied Scientific Research TNO, Report R11506, 27 November 2015, Leiden, The Netherlands https://repository.tno.nl/islandora/object/uuid:f67c5ad8-1c27-4768-97b9-3d842f228379; Zwetsloot, G.I.J.M., Kines, P., Ruotsala, R., Drupsteen, L., Merivirta, M.L., Bezemer, R.A., (2017) The importance of commitment, communication, culture and learning for the implementation of the zero accident vision in 27 companies in Europe, *Safety Science*, 96, 22–32. Doi: 10.1016/j.ssci.2017.03.001.
42 Beer, M., (2009) *High Commitment, High Performance – How to Build a Resilient Organization for Sustained Advantage.* San Francisco: Jossey–Bass; Walton R.E., (1985) From control to commitment in the workplace. *Harvard Business Review*, 63 (2), 77–84.
43 Edwards, M., Jabs, L.B., (2009) When safety culture backfires: Unintended consequences of half-shared governance in a high tech workplace, *Social Science Journal,* 46, 707–723. Doi: 10.1016/j.soscij.2009.05.007.

4 Vision Zero

Underlying Concepts and Theories

In the former chapter, it was shown that Vision Zero is more than the ambition to realize a challenging long-term goal. It is based on SHW as human and social values, on synergies between the three aspects of SHW and with business processes. It is based on values and ratios. It requires a long-term strategy, visible leadership commitment, as well as the commitment of the majority of the workforce. Vision Zero requires processes of learning and innovating to zero. In this chapter there is more in-depth focus on a number of concepts and theories that are important for understanding and implementing Vision Zero and for the process of continual improvement.

4.1 VALUES THAT SUPPORT VISION ZERO

In the former chapters, the value aspect of safety, health and wellbeing was already shown to be important for organizations, their leaders and the majority of the workforce. Besides SHW as values in themselves, there are other human and social values that support prevention and SHW.[1]

Values are an important but often unnoticed factor in organizational and societal cultures. Values give direction on what is really valuable and desirable. It is therefore easy to understand why organizations pay increasing attention to values: they co-determine their functioning. By defining and sharing core values, companies give direction to their development and actions and can give meaning to the company's existence and its value for society. Values serve as a compass for strategic decisions. They are important for the *purpose* of organizations and for creating *meaningful work* for their members.

Michael Beer[2] concludes that 'employees, who are psychologically aligned with the mission and the values of the organization, are intrinsically motivated and generate high performances'.[3] This implies that when an organization and its members share SHW supporting values, it will not only strengthen SHW but also improve work performance and productivity. Values become ways of reasoning as well as moral obligations, and in this way guide organizational practices and the behaviour of senior leaders, supervisors and workers. They become part of the organizational identity.

Values can help to prioritize issues, but a value is not a priority. Priorities tend to change quickly in changing circumstances. Values that are internalized are more

DOI: 10.1201/9781003621225-4

stable than corporate structures, management systems or priorities. That is especially relevant in a world of change. Values can therefore serve as a compass in times of change. Contrastingly, values that are only espoused have little importance and can easily be disregarded.

Core values often emphasize the importance of business goals (e.g., innovation, reliability), sometimes more societal issues (e.g., sustainability) or SHW, but they can also include other human and social values. A set of core values only has a positive impact on SHW when some of the values are social values that resonate with the values of the people at work. That is why business values should be combined with the values of the working population. For a prevention culture, it is important that prevention is integrated into the values, mindsets and actions of managers and employees.

4.1.1 The SHW Supporting Values

In a comprehensive literature review, seven (core) values were identified that support SHW.[4] Three values are relevant for 'being' (interconnectedness, participation and trust), two for 'doing' (justice and responsibility) and two for 'becoming' (growth and development and resilience). These values are important for SHW performance and the development of a prevention culture.

Interconnectedness is a fundamental value because human beings are not only individuals but also social beings. Bonding is a basic need for each individual, and affective involvement is important for every individual. Personal identity is closely related to the social environment, including the work environment. Similarly, organizations are not only economic entities but also social communities, with both managers and workers as members. Good teamwork and social support from colleagues and leaders show the worker that (s)he is not alone. This is directly important for wellbeing, for engagement and indirectly also for safety and health and the business. For many people, the social contacts with their colleagues contribute to the liking of their job.

Participation implies the active involvement of workers. Every individual has an interest in SHW. A safe and healthy working environment is now globally recognized as a fundamental labour right. That right requires active participation in the form of direct participation complemented by representative participation. Representative participation is a legal requirement in most countries. Nationally and internationally, tripartite organizations (e.g., ILO) determine SHW policies, also a form of representative participation at the international level.

Workers often have useful ideas on how to improve or facilitate their work. Based on their work experience, workers have tacit knowledge and know what can work in practice. When they have a say in how their work is organized, they can significantly improve the organization and planning of work or the (re)design of the working environment including tools to work with. Workers should not only be protected – that would imply only a passive role for them.

Trust starts with mutually respecting others and requires the absence of fear. Trust between management and workers is only a reality when it is mutual: when workers trust their managers (of any hierarchical level) and each of these managers trusts the people they are leading. Trust between co-workers is important for good teamwork. In case of distrust, communication will be hampered as people are likely to suspect negative reactions or hidden agendas. Then, messages will be taken less seriously and have little impact. People will be reluctant to report incidents or errors, making managers uninformed about practical problems and hampering SHW policies to be effective. Trust is also important for the implementation of SHW interventions (less resistance to change). Transparency is important for building trust, but private data needs to be treated confidentially. In case it is unclear why things are confidential that potentially undermines trust.

Managers may distrust their people or be anxious, which may trigger them to perceive critical questions or suggestions for improvement as a threat to their position and status and to react authoritarian. When managers do not take critical or warning signals seriously, the effect is that people stop making critical questions or suggestions and stop reporting critical situations. This is the condition for developing a tunnel vision and is known to turn workers and other stakeholders into passive bystanders. A recipe for disasters.[5]

Another type of trust is being confident that OSH management systems, safety rules, other parts of the organization and other people are functioning well. Too much of that type of trust can have negative effects on SHW. It may trigger complacency, lower alertness on SHW risks and less critical reflection in unclear or ambiguous situations. Similarly, when managers have too much trust that individuals can take care of themselves, e.g., when they suffer from stress or have to solve problems in their work, that can also work out negatively. These aspects of trust require a good balance between confidence, alertness, critical reflection and social support.

Justice. James Reason emphasized the importance of a *just culture* for safety: an atmosphere of trust that encourages people to deliver SHW-relevant information and where everybody knows what is acceptable or unacceptable behaviour.[6] Justice and reliable information, even if it is bad news, generate credibility and confidence in management. That is just as relevant for health and wellbeing as for safety.

In the literature, a just culture is seen as an important prerequisite for open and honest communication, e.g., for reporting incidents and near misses. Justice and trust are important preconditions for sharing sensitive, often personal, information,[7] such as feelings of discomfort, personal problems and doubts. Blaming the victim is the most well-known expression of the absence of justice. Injustice immediately turns trust into distrust.

Procedural and distributive justice are regarded as two distinct determinants of mental health at work. If people are not treated in a just way, they can easily become frustrated or stressed and in the longer run develop burnout or start looking for another job.

Responsibility. Responsibility implies that people and organizations take responsibility not only for the accomplishment of their tasks but also for unintended consequences thereof. Not only for their own safety and health but also for the SHW of others (colleagues, trainees, business partners, etc.). It implies the recognition that prevention is a joint responsibility for all. A responsibility for (pro)actively dealing with risks, doubts, ambiguities and involving others timely. Taking responsibility is an essential aspect of leadership and proactivity. Taking responsibility is important for *ownership*: recognizing that you have to play an active role in solving certain SHW problems. The opposite of taking responsibility is being complacent. Complacency is known to be a determinant of accidents as well as organizational health and wellbeing issues.

Development and Growth. Development and growth are important for organizations as well as for individuals. For organizations, it includes the challenge of growth and innovation. Smart organizations have an interest in the growth and development of their personnel as that is essential for innovation and positive change.

For individuals, development and growth means developing new competences and developing self-confidence. It is important to create the conditions for *learning on the job*, i.e., for a work organization that fosters learning and development. That requires a degree of autonomy and control of the work processes. Individuals have a fundamental need for self-realization and self-actualization. Individual development and growth imply learning and underlie many processes of continual improvement. It implies that the focus should not only be on (existing) problems but even more on opportunities for improvement. Especially for more fundamental solutions, it is often necessary to think out of the box and to critically reflect on prevailing mental models.

Resilience. Resilience implies the ability to adapt to changing circumstances, to remain confident and vigorous when confronted with difficulties and to recover quickly thereafter,[8] see Section 2.3.2. It is important for the functioning and survival of organizations and individuals in times of adverse conditions. Sometimes adverse conditions cannot be foreseen; then it becomes necessary to *manage the unexpected.*[9] Psychological alignment of individual and organizational values and goals is an important condition for organizational resilience and individual wellbeing, as well as for any commitment strategy.[10]

4.2 ALIGNING AND SHARING VALUES

4.2.1 Aligning Values

Each of the values mentioned above supports SHW. They are complementary but not fully independent and do not generate a positive impact in the same way. The relationship between justice and trust was already mentioned. Other examples are the link between participation and development and growth, or between interconnectedness and taking responsibility for the SHW of others. There is considerable potential for synergy between the seven values. In practice, this is strongly influenced by the way the values are operationalized, understood and aligned or shared.

The seven core values that support SHW also contribute to business success. They are important for being an employer of choice, for having and retaining motivated and loyal employees, and for open and honest communication. For individuals, adopting organizational values can contribute to being a valued and respected member of the organizational community.

Aligning organizational values and goals with those of the employees is a prerequisite for the internalization of values into the organizational culture. The aligning process implies that the values that senior management sees as important for the organization (often business-related values only) are aligned with the values that the workforce regards important (often including social values, e.g., SHW and some of the supporting values mentioned in the preceding section). In this way, a common value basis can be generated for the organization as a community. Besides the common basis, people in the organization will always adhere to a variety of personal and group values.

As a first step in the alignment process, the values of both sides of the organization have to be made explicit. This may require considerable dialogues in either group. Then, joint dialogues of management and the workforce are required, e.g., in the form of a 'World Café'.[11] It presupposes an open attitude of senior management as well as the majority of the workforce. As Peter Senge said: 'The alignment of values is not a rational process but requires that managers and workers engage with their hearts. It requires them to "become in touch with their humanity and their values" '.[12] Examples of what a value could mean in terms of practical consequences have to be discussed as well. Of course, both sides have to seriously consider the values proposed by the other side. It is best when a mix of values proposed by senior management and those by the (representatives of) the workforce will come out as the organizational values as a result of the alignment process (see Figure 4.1).

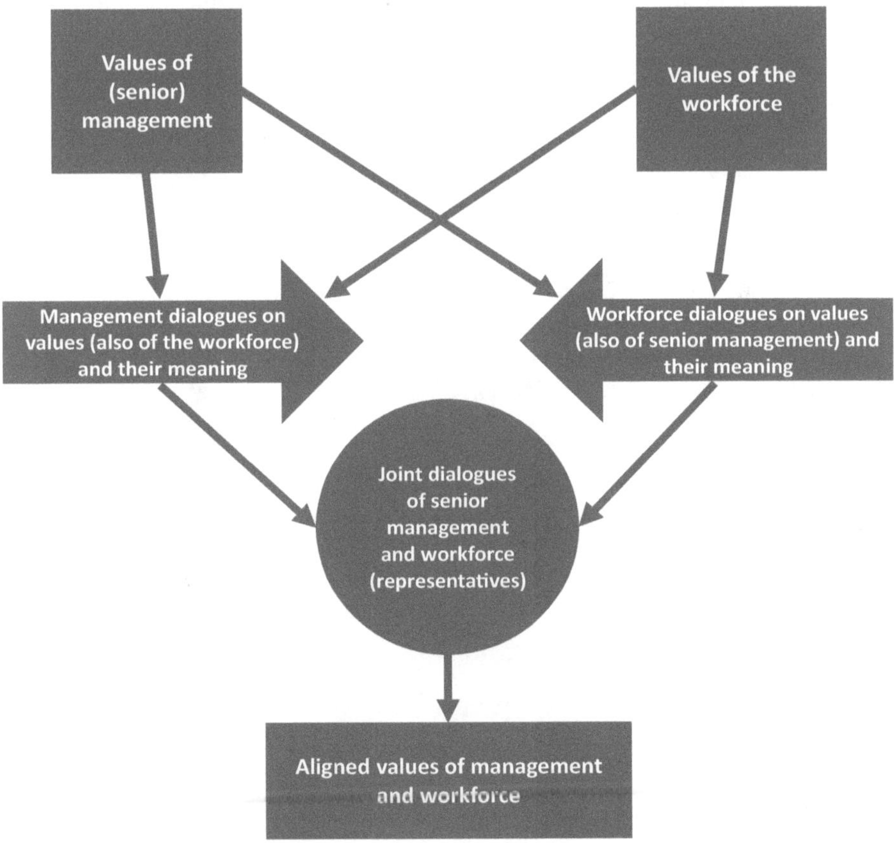

FIGURE 4.1 The process of aligning values of senior management and the workforce.

4.2.2 Sharing Values and Giving Them Concrete Meaning

The aligned values of senior management and (representatives of) the workforce form a good starting point for sharing those values with all members of the organization, giving them concrete meaning and integrating them into the policies and culture of the organization. This will require alignment with the personal values of most members of the organization. To further develop aligned values into shared values that are internalized by managers and the workforce requires further dialogues and participatory processes.

In the process of aligning and sharing values, it is important (1) to give further concrete meaning to the values and (2) to create a participative process to share them with all members of the workforce (see Figure 4.2). The process of sharing values with a majority of the workforce will require time for dialogues and active participation. Shared values cannot simply be planned and deployed. It is not a matter of simply making a statement or implementing an instrument. To share values, they have to be adopted and *lived* by most members of the organization and confirmed in social interactions with other members before they are really internalized as *shared values*.

Values are multi-layered, just like organizational culture. It must become clear what the values mean in practical situations, for (senior) leaders as well as the workforce. To this end, the values have to be operationalized for the specific context,

FIGURE 4.2 The process of sharing values and giving them concrete meaning.

including the social, cultural and economic context. That can be done by using associated keywords and examples of desirable or undesirable behaviour.

Table 4.1 gives a non-exhaustive list of relevant keywords and associated desirable behaviours for the seven values mentioned earlier.

TABLE 4.1
Examples of key words and desirable behaviours relevant for the operationalization of the SHW supporting values

Core Values	Associated Keywords	Examples of Desirable Behaviour
Interconnectedness	Collaboration Collegiality Community Meaningful (open and trusted) relationships Social cohesion Social support Social capital Team work	Be empathic Reward taking initiatives Show leadership (also bottom-up) Support your colleagues Treat colleagues in the way you want to be treated yourself
Participation	Autonomy Empowerment Engagement Everybody counts Ownership Self-organization Social inclusion Tacit knowledge	Involve people in decisions that impact their life or work Recognize practical expertise Making use of the drives and tacit knowledge of workers Focus on people's abilities, not on their disabilities
Trust	Consistency Dignity Fair and open communication/ Transparency Integrity Loyalty Supportive feedback Psychological safety Reciprocity Respect	Avoid double standards and hidden agendas Be consistent in behaviour Give constructive feedback Practice what you preach – Be the example Respect each and every individual Create an atmosphere wherein people do not hesitate to report problems, incidents or errors
Justice	Decent work Diversity (No discrimination) Effort and reward balance Equity Fairness Fair contracts Fair pay Integrity No fear or anxiety	Avoid discrimination and favouritism Balance trust, accountability and learning Regard incidents and mistakes as opportunities for learning Treat people the way you want to be treated Use power carefully

(continued)

TABLE 4.1 (Continued)
Examples of key words and desirable behaviours relevant for the operationalization of the SHW supporting values

Core Values	Associated Keywords	Examples of Desirable Behaviour
Responsibility	After care Anticipation Dealing with problems Early warnings Prevention Proactivity Taking 'Ownership'	Maintain work–life balance Share reliable information Take a long-term perspective Take care of your colleagues and their families Take responsibility when SHW (or other) warnings are recognized or when problems occur
Development and growth	Constructive feedback Creativity Future orientation Informedness Leadership Learning Long-term perspective Meaningful work Purpose Reflexivity Self-realization and self-actualization	Develop new skills and competences Give timely and constructive feedback and be willing to receive and value it Learning on the job Make sure all jobs (and most tasks) are meaningful Show and develop leadership (at all hierarchical levels) Use problems and difficulties as opportunities for learning
Resilience	Adaptivity Anticipation Confidence Flexibility Organizational-mindfulness Recovery Remain confident Sense-making Social capital	Adapt timely to changing circumstances Anticipate the unexpected Maintain health and employability Organize dialogues on the meaning of unexpected phenomena (problems) and warning signals of any kind

Source: Elaborating on Zwetsloot, G.I.J.M., van Scheppingen, A.R., Bos, E.H., Dijkman, A., Starren, A., (2013) The core values that support health, safety and well-being at work, *Safety & Health at Work*, 4, 187–196. Doi: 10.1016/j.shaw/2013.10.001; Zwetsloot, G.I.J.M., (2019) Shared values for health, safety and well-being at work, in: Burke, R.J., Richardson, A.M., (eds), *Creating Psychologically Healthy Workplaces*, Cheltenham: Edward Elgar. pp. 91–111. Doi: 10.4337/9781788113427.00011.

A selection of two or three keywords and two examples of desirable behaviours per value will probably be sufficient. The selection of keywords and examples of desirable (and undesirable) behaviour should be tailor-made to the context of any company. In this way, the operationalized values will be unique for the organization.

The processes of aligning and sharing values imply that the managers and workers become increasingly intrinsically motivated to pursue the shared organizational values and deliver high performance. The shared values and their concrete meaning can serve as the ethical compass for all members of the organization. That will contribute to being inspired, motivated and engaged and to experiencing work as *meaningful work*.[13] Meaningful work keeps people motivated and makes them proud of their job, and the products or services they provide. This strengthens their engagement/commitment. It also helps them to identify themselves with the organization and its brands. It means that potential value conflicts between personal values and work will be rather exceptional. In this way, the shared values become organizational resources that complement job and personal resources in generating engagement.

It is essential that the commitment to the shared values by top management is genuine and that they are willing to *live and work* according to the values. Commitment cannot be developed in a rational way only – exemplary behaviour and leadership are essential. For credibility in the eyes of the work force, it is essential to show that the values, once adopted, are not optional. SHW investments in hard times will be perceived as signs of real commitment that strengthen the credibility of the organizational values for longer periods of time. It is also important that the shared values are communicated regularly, through formal company communication as well as through visible leadership commitment and informal dialogues. Dialogues and leadership are especially important in case dilemmas turn up in practical situations. It is also important to share the values early in the process of selecting and hiring new personnel. It can serve as selection criteria not only for the company but also for the candidate(s) for the new job.

Sharing organizational and employee values often results in the organization becoming an 'employer of choice'. Most people prefer a job wherein they can recognize a number of their own values and work that is experienced as meaningful. The purpose of work is also determined by the broader results of the work, i.e., the social and societal impact of what is done (e.g., contributing to people's health or to sustainability).

4.2.3 The Broader Meaning of the SHW Supporting Values

Values that are shared and internalized have a relevance that does not stop at the fence of a production plant or the walls of a workplace; such values have wider implications. Only a few examples are mentioned here: For an organization, the values of interconnectedness and responsibility can be relevant for valuing and respecting human rights, the company's neighbours and the rights of future generations. For individual managers or workers, it may impact their activities at home and in leisure. Participation is not only relevant for employees; for organizations, it can mean involvement and cooperation with external stakeholders (a key issue in CSR). For workers, it may imply a greater willingness to participate in public issues and voluntary work. Trust is not only a basic requirement for good communication and cooperation at work but also for cooperation externally, e.g., in public–private partnerships. The values of justice and responsibility are clearly relevant for fair business practices, social responsibility and for open and honest communication with stakeholders. On the individual level, they are meaningful for voluntary work or donating to charities. Growth and development and resilience are relevant for economic development, for society as a whole and for personal development.

4.3 OWNERSHIP, INTRINSIC MOTIVATION AND PURPOSE

What drives managers and workers to work in a safe and healthy way that also contributes to their wellbeing? How can they be motivated to actively commit themselves to Vision Zero? People can be motivated in many ways and for many different reasons, but some reasons and ways are more useful for Vision Zero than others.

First of all, it is important that contributing to SHW is an integrated part of work. Otherwise, SHW can easily be perceived as an added-on, as no more than a secondary issue. In case of production pressure or crises, priority will then be given to production activities, not to SHW. When SHW are integrated aspects of work this means the motivation for working and for contributing to SHW coincide. This brings us to the purpose of work.

4.3.1 Internal Motivation and Purpose

Most people start working because they need money. It is important that people are paid sufficiently for their work to have a decent life with their families and perhaps enjoy some luxury. But earning money is not the purpose of work. Also, for companies it is an outdated mental model that their only goal is earning money. Fortunately, organizations increasingly realize that positive social and ecological impacts are just as important as being profitable. Earning money is not the purpose of life. Arie de Geus, a higher Shell official, said in this respect: ‘Businesses need profits in the same way that any living being needs oxygen: we need to breath in order to live, but we do not live in order to breath’.[14] Many people working in health care are motivated for their work because they experience it as heart-warming to care for people who need help. Many teachers in elementary schools love to work with children. Many people working for companies that are frontrunners in environmental sustainability are deeply motivated because they believe their work makes them contribute to a more sustainable world (also for their children and grandchildren). And many SHW professionals are motivated for their work because they can tangibly contribute to the safety health and wellbeing of people through their work. The common denominator of these examples is that work can be socially or ecologically meaningful. Work can also be meaningful because it has an appeal of novelty and play, because it is inherently interesting, has an appeal of exploration and learning or has aesthetic value.[15] Two complementary factors known to be motivating for non-paid work are also relevant: getting to know new types of people (i.e., increasing social capital) and the opportunity to acquire new skills and competences[16] – in other words, when it contributes to personal development.

Intrinsic motivation means that people are motivated by the inherent satisfaction of doing the activity itself[17] – the activity is its own reward. When people perform activities they are intrinsically motivated for, they work energized and persistent while it contributes to their wellbeing, self-esteem, confidence and vitality.[18] For the organization and the people concerned, behaviour based on intrinsic motivation is a renewable resource of mental energy.[19]

But for many people, intrinsic motivation is only scarcely found in their work: they do not experience their work as meaningful. There is overwhelming evidence that deadlines, directives, financial rewards and sanctions, production pressure and imposed goals affect intrinsic motivation. In this way, the organizational conditions can easily affect or even destroy the intrinsic motivation of people. In many organizations, intrinsic motivation is not encouraged but unintendedly discouraged. They focus on external motivators which affect wellbeing and do not trigger intrinsic motivation.

For SHW, it is most desirable when all managers and workers are self-motivated to contribute to excellent SHW. Relevant literature[20] refers to the importance of intrinsic motivation for a proactive or generative safety culture. Others mention the importance of intrinsic motivation for safety culture to ensure that the workforce is working safely also *when nobody is watching*, i.e., as a kind of internal control mechanism. Indeed, it is important that people are motivated by themselves for being proactive, creative and for feeling 'ownership' (i.e., feeling personally responsible and motivated). These are all self-directed processes.

Intrinsic motivation is also important for managing cultural change. Giving resources and encouraging internally motivated persons to act in a way they are already motivated for can inspire and motivate others. They can show exemplary behaviour and exude confidence and enthusiasm. That is often more convincing than rational argumentation by management or SHW staff.

The opposite of intrinsic motivation is extrinsic motivation; in that case, the rewards stem from outside the activity, e.g., financial rewards or status. The opposite of meaningful work is a bullshit job[21] – a job that does not contribute in any way to a better world.

The distinction between intrinsic and extrinsic motivation is not a matter of white and black. There are also lighter or darker greys: forms of motivation that have a mix of internal and external components. Two of the lighter greys concern forms of motivation wherein most of the motivation comes from the self: identification and integration.

When the external conditions and the social environment are aligned with the self, people may value a goal (e.g., zero accidents or zero work-related illnesses) in such a way that it is important for them personally. It is a conscious decision to adopt a goal as personally relevant. They then *identify* themselves with the goal – they feel ownership – and often they identify themselves with the organization or brand that is striving to achieve that goal. This makes them proud of the work as professionals. That is also important for the retention of high-quality personnel. When contributing to SHW, excellence is seen as part of being a good professional, of mastering one's job or of being a member of an excellent team that identification implies a strong and stable internal motivation. SHW is then much more than an externally imposed priority.[22]

This is also confirmed by the literature on *work engagement*. Engaged employees put much effort into their work because they identify with it.[23] They perceive their work as challenging and significant. They work with enthusiasm and a high level of energy and strongly identify themselves with their work. To facilitate work engagement, it is important to achieve a good balance between job demands and organizational, job and personal resources. Autonomy/job control and social support by leaders and co-workers are key resources any organization can influence. Redesign of the work organization may be needed to create conditions for more engagement and wellbeing. It is also important to avoid or reduce activities that hinder engagement, e.g., bureaucracy, role conflicts or hassles.

When the external demands are not only adopted consciously but are fully assimilated and brought into congruence with one's values and needs, it results in *integrated* motivation. That is more than a conscious and rational process, the personal aspirations are then fully aligned with the demands of the work processes and the social environment.

Forms of external motivation also generate action, but then it is often accompanied by feelings of anxiety (lack of wellbeing) and poorer coping with failures. This is often the case when there is a strong organizational or personal emphasis on compliance, external rewards and sanctions. External motivation is easier to create but disappears quickly when the external triggers are no longer present.

4.3.2 Conducing Ownership and Intrinsic Motivation

Though the three forms of internal motivation cannot be generated externally, they can be enabled by organizations and their leaders. Motivation to contribute presupposes that people are actively involved. People feel ownership for preventive measures when they are involved or consulted during the development of those measures. Early participation is important for the development of ownership. This is also true for the investigation of unexpected events: when managers and workers are actively involved in the investigation, the outcomes and learnings will most likely be owned by them. We already saw that participation is an SHW supporting value. Interestingly, the Dutch labour inspection now applies this principle to increase the motivation and learning of companies after serious (but not fatal) accidents: companies are invited to carry out the accident investigation themselves (with some guidelines and supervision from the inspection), which is essential for developing ownership. When they do the investigation seriously, the fines will be significantly lower or even skipped at all. The companies are expected to own the conclusions and recommendations of their investigations. That increases the likelihood that the company learns from the accident and invests in SHW improvement – the inspection supervises the implementation of the lessons learned.[24]

To conduct intrinsic motivation, the combination of three factors – representing three basic human needs – is essential: a sense of *autonomy*, *competence* and *relatedness*. According to the self-determination theory, these three factors are actually basic human needs, and their fulfilment contributes strongly to internal forms of motivation. The challenge is therefore to develop contexts (work organization, leadership, culture) that are supportive of these three factors. The three factors are therefore explained in a bit more detail. The term internal motivation is used in this book for motivation that is intrinsic, integrated or based on identification. Then, the locus of control is internal, and people feel ownership.

Autonomy. People want control over their activities; they do not want to be controlled. People want to develop themselves. They want to have a say about their own life as well as their work. They are usually not against change as such, but they do not want to be changed. Ownership presupposes a fair amount of 'control' over the activities at work. On what people do, when they do it, how they do it and whom they do it with. Having control is an essential condition for the (development of) ownership. The work organization and the employment contract determine how much control people have, and thereby the degree of autonomy. It requires leaders that empower their people to organize or adapt the work based on their competence and experience. Autonomy does not mean to be independent, detached or selfish, but it rather means the feeling of volition that can accompany any act, whether dependent or independent, collectivist or individualist.[25]

Competence. For internal motivation and ownership, it is important that people experience self-efficacy, i.e., that they feel they are competent. It is important that the organization recognizes the experience and tacit knowledge of the workers and defer to their expertise. Learning and personal development and growth (already mentioned as supporting values) are important for competence; learning on the job can be supported by an adequate work organization. According to Pink, there is often a mismatch between what people must do and what people can do.[26] When what they must do exceeds their capabilities, the result of the work is of lower quality while for the people involved it generates anxiety. For leaders, it is important to act in ways that people feel their expertise is recognized as important and can be further developed. For instance, to encourage employees to voice concerns when they see a problem, or in case of discussions to use open questions and focus on good listening, instead of explaining and telling.

Relatedness is about belonging to a group and being respected by and connected with your colleagues and your supervisor (comparable with the supporting value of interconnectedness). It presupposes mutual trust and justice (again supporting values). It also implies that there is some time to develop and maintain social contacts and get to know each other; a high turnover of staff or many temporary workers can undermine the team spirit. *Social support* is another key factor in enabling internal motivation: the experience that you are not alone when confronted with difficulties or dilemmas. That you can count on the support of your co-workers and supervisor, even in case you have made a mistake. Constructive feedback contributes to social support. As mentioned earlier, social support is also a key factor for work engagement. Note that relatedness and social support are both expressions of the SHW supporting value of *interconnectedness*.

For leaders, it is important to see the employees as individual people, as partners, not just as human resources. With people you have personal relationships, resources are just something to use. Good practices are to engage in dialogue and debate, to give social support, to stimulate participation and to ensure that incidents and problems are investigated without fear or blame.

4.4 DOING THE RIGHT THINGS RIGHT

4.4.1 Doing Things Right

The goal of Vision Zero implies that the quality principle of *doing the right things right from the start* is a leading principle for the business as well as SHW activities. *Doing things right* implies that production activities and work are well prepared before the execution starts. It implies timely anticipation and alertness on potential problems, especially in the early stages of research, development, design, procurement and planning. It also requires ensuring the capability and competence to take adequate action whenever a problem is identified. Decisions in the early stages determine to a high extent whether work during production can be done right, i.e., without generating (SHW) problems that are foreseeable. The principle of *doing things right from the start*, i.e., already the first time, is also relevant for being proactive.

> The principle of *doing things right* (from the start) is not the same as the focus on *how things go right* in Safety-II. Hollnagel's *how things go right* focuses on the stage of production and normal work; it is not possible to operationalize it in the design stage[27] to anticipate future (SHW) problems in the early stages. The Safety-II principle focuses on dealing adequately with variability during production, whereby the processes that generate that variability are taken for granted.

Doing things right is a rational objective that is well known and broadly accepted. Educational institutions, from basic education to universities, promote the principle. We are trained to think rationally, and for many rational issues there is usually only one valid best solution. Many industries are therefore aiming to implement *best practices. Doing things right from the start* has the potential to greatly reduce the number of problems and the need for corrective action. Economically, it makes sense: it has the potential to save a lot of costs. Nevertheless, it is often regarded as neither realistic nor feasible. It usually requires some extra time for reflection and exploration of options in the first phase of decision-making. Unfortunately, the extra costs for careful decision-making are easily identifiable and have to be made at the beginning of the process, while the benefits may be partly intangible and are only achieved later. Organizations often focus on short-term impacts and hurry to solve a problem or start a new activity. It can then hardly be surprising that activities often *go wrong from the start.*[28]

4.4.2 Doing the Right Things

Doing things right is, however, not always sufficient for an excellent result. A quote from Juran, a thought leader in total quality management from the twentieth century, clarifies this: 'If the goals are poorly chosen, the planning will be done to reach the wrong goals. We shall be "doing things right" but not "doing the right things"'.[29]

The two dimensions of business and SHW excellence	**Doing the right things**	**Doing the wrong things**
Doing things right	Doing the right things right **(excellence)**	Doing the wrong things right **(major problems)**
Doing things wrong	Doing the right things wrong **(minor problems)**	Doing the wrong things wrong **(major problems)**

FIGURE 4.3 The two dimensions of (business and SHW) excellence.

Juran's quote clarifies that *doing the right things* is independent of *doing things right.* The two dimensions can be represented in a two-by-two matrix[30] (see Figure 4.3).

Doing the right things implies that the goals are aligned with the organizational core values and contribute to the strategic choices of an organization or an individual (manager, worker, etc.). For people it is about purpose, about giving meaning to their life and work. *Doing the right things* is therefore – in principle – important for any organization and individual. Different from the principle of *doing things right*, there is usually not one best option for *doing the right thing* because there is never one best answer to questions like: 'What is the best investment this business can make?' or 'What is the best way we can improve the wellbeing of our employees'. What is right depends on the values of the persons and organization involved and is often influenced by the context. Both externally and internally, organizations are confronted with a range of stakeholders, each having their own specific values and interests. The values and interests of stakeholders form an important context factor that might influence what is the right thing to do for organizations. Consequently, there is not one best way for all companies to implement Vision Zero. It is better to speak about good practices rather than one best practice.

It is important to understand that *doing the wrong things right* is much worse than *doing the right things wrong*. Both are OK in one dimension and not OK in another, but the two dimensions are not of equal weight. For organizations, it is everyday practice to find out that there are problems or opportunities for improvement. But it is much more difficult to realize that they are doing the wrong things. At first sight, it seems attractive to make the step from doing wrong things wrong to *doing wrong things right.* But this will bring you exactly to the situation Juran warned of. *Doing the wrong things right* (the upper right quarter in Figure 4.3) implies that the organization is pursuing aims that are not OK (e.g., not compatible with their core values and identity, or wrongly operationalized goals). This will lead to persistent problems and friction between these activities and the organizational strategy. A normal problem-solving process will not help. The organization is on the wrong track! At the end of the day, this usually means that the organization must stop these activities and start all over to do the right thing. When the problem is that *the right things* are done *wrongly* (left below in Figure 4.3), the activities can usually rather easily be improved, good problem-solving is then sufficient to solve relatively minor problems and to achieve excellence.

Doing the right things is relevant for every decision or action at any organizational level. But in organizations, the decisions and actions of those at the top have the greatest impact. It is extremely important to *do the right things* in the boardroom. But the principle is also vital for operational activities (and being a good professional). It is relevant for decisions and activities at any stage in the life cycle of products, services and processes. Most important are, however, the decisions made at the strategic level, and in the phases of design and planning.

Doing the *wrong* things may result when values other than SHW (e.g., productivity) are regarded as more important. That leads to half-hearted SHW goals or half-hearted support to achieve the (SHW) goals (e.g., not sufficient people or money or time available). Doing the wrong things can also result from the wrong operationalization of the right goals. Consider the English saying *Pounds foolish, penny wise*. Being penny wise can be pursuing the right goal, but the approach to saving money does not make much sense if you are pounds foolish. Similarly, the Dutch saying *Mopping with the tap open* clarifies that the context factors do make a difference and may change sound activities like good mopping into a useless and hopeless activities. See Table 4.2 for examples of doing the wrong things.

Each of the above-given examples of *doing the wrong things* can be done *right* or *wrong*. But even when they are done rightly the results will be disappointing. The common characteristic of *doing the wrong things* is that they do not have the potential to be successful.

TABLE 4.2
Examples of doing the wrong things

Acting in conflict with the organization's core values or strategy
Accepting trade-offs in SHW goal setting
Considering the system too narrowly for taking into account all relevant determinants of SHW risks
Choosing a technology that is not able to deliver the performance (e.g., precision) aimed for
Involving people who are not sufficiently competent to realize the aims
Aiming at short-term improvements without taking into account the long-term impacts thereof
Stopping accident investigations when a human error is identified as a cause
Focusing on solutions on paper that will have a minimal impact on actual practices. The administrative aspects should never become an aim in itself
Fighting symptoms instead of dealing with underlying causes
Focus on the behaviour of individuals without considering organizational factors
Taking action in a situation that is objectively safe and healthy but is *perceived* as unsafe or unhealthy
Creating a situation that is *perceived* as safe and healthy but that objectively is not safe and healthy

Mental models play a role in goal determination as well as in problem analysis and choice of approach. Wrong goals or inappropriate approaches can be chosen because the decision-making process is biased, e.g., in cases of wishful thinking, jumping to conclusions, hindsight bias, group thinking, etc.[31]

But nobody is perfect, no organization is perfect. Perfectionism often leads to high levels of control and often to stress. We live in a world of *good enough*. This contributes to not only the flexibility but also the complexity in organizations. It is important to distinguish between the conditions where perfection is really needed due to high SHW risks with potentially irreversible impacts (death, disability, etc.) and situations where good enough is to be preferred. In the former, the so-called *Red Rules* are strictly mandatory, and in the latter more loose forms of risk control can be preferred and rules do express behavioural expectations.[32]

4.5 PROACTIVITY

Vision Zero cannot be realized by merely focusing on compliance, or by being reactive. Instead, it is important to be proactive and forward-looking. Let us consider the concept of proactivity and its implications for companies that adopt Vision Zero.

The term proactive refers to three types of self-initiated efforts to bring about change to achieve a different future. It refers to *making things happen*, *anticipating and preventing problems* and *seizing opportunities for improvement*.[33] Proactivity is a goal-driven process involving both goal generation and goal striving. It is motivated, conscious and goal-directed. Proactive people may introduce SHW improvements because they enjoy their work greatly and see it as improving their work further, and/or because an excellent SHW performance is regarded as an important aspect of their professionality and their professional identity.

Making things happen implies taking the initiatives to generate something new that is expected to contribute to a better future, in our case a future free of work-related accidents and ill health. Making things happen is a self-starting process based on ownership and personal initiative; it is change-oriented and future-focused.[34] Like intrinsic motivation, it cannot be generated externally, but indirectly it can be promoted and encouraged by inspiring and empowering people, and by creating conditions that are conducive to proactivity.

Vision has been identified as a key element of transformational leadership for proactivity. Vision Zero is an inspiring (long-term) goal for many organizations and individuals. In this way, Vision Zero provides a motivational force for proactive action.[35] Proactive goals can be quite ambitious. They are often challenging and require perseverance.[36] This is especially relevant for long-term-oriented proactive goals like Vision Zero. Therefore, it is important to clearly explain and share the ‘why aspects’ of Vision Zero. Though Vision Zero and its goals are rather abstract and high level and concern the company as a whole, most workers recognize the importance of each of the three SHW dimensions for themselves as well as for their family life. It is important to emphasize that Vision Zero is not only an issue of organizational performance but also of individual and collective commitment and learning.

Early anticipation to prevent problems from arising requires early recognition of potential SHW problems. When SHW is anchored and integrated into the business processes, SHW will be considered in case of changes in production processes or organizational arrangements. The SHW aspects are then taken into account in the earliest stages of the process, preferably with the participation of workers' representatives. When SHW is integrated into the development and design of new business models, new business activities, workplaces, products, production processes, the planning of production activities, projects and work processes, future SHW problems can be prevented. Anticipation also implies learning from undesirable or unexpected events like incidents (accidents, near misses, cases of work-related ill health and burnouts, complaints about bullying, discrimination, etc.) to prevent reoccurrence. In many companies, it is regarded as important to learn from (safety) incidents, while health and wellbeing events are often wrongly regarded as personal issues that are not relevant for organizational learning.

Seizing opportunities includes learning from other organizations, in the same sector as well as cross-sectional, such as in national Vision Zero networks.[37] To learn from innovations developed by others, and from developments in the state of the art knowledge. To use suggestions from workers – they often have tacit and practical knowledge to improve their work. It is also important to learn from situations wherein things went remarkably better than expected.

With a proactive goal in mind, it is possible to recognize opportunities that can contribute to the realization of that goal. The opportunities may arrive unexpectedly and may come in disguise: at first sight, they may appear as deviations and raise questions about what is happening. They may appear as problems or as warning signs for upcoming problems. But what appears as a problematic may trigger processes of reflection and learning leading to new insights which creates opportunities for improvements. Proactivity helps to deal adequately with unexpected events.

In the theory of high-reliability organizations, the importance of recognizing weak signals is emphasized[38] (see Section 9.4). Such signals may relate to technology, organization or people and their behaviour. Proactively using early warnings and potential problems as opportunities for improvement is strongly influenced by the organizational culture. The term proactive culture is frequently used to refer to a high level of safety culture maturity.[39] The term a 'generative safety culture' is used by Hudson and colleagues for the highest level of safety culture maturity, but in fact it is referring to proactive behaviour of the 'making it happen' type.

The organization of work is important for proactive work behaviour. Enriched jobs (i.e., jobs that include opportunities for controlling the work process) create conditions under which individuals experience enjoyment and flow and are thus intrinsically motivated to be proactive in their work.[40] Job autonomy and having control predict proactive work behaviours including personal initiative, idea implementation and suggestions for improvement, while lack of job control inhibits proactivity. Social support (by co-workers and managers) positively relates to various proactive behaviours at work.

It is impossible to force people to be proactive or to trigger it externally. Prescriptions will not work: proactivity is a self-initiated effort; *ownership* is crucial.

Promising financial benefits in case of proactive behaviour will not work either, though it might be useful to reward proactive behaviour afterwards, as an acknowledgement and encouragement. Proactivity can be encouraged by inspiring leadership, empowering people and a commitment strategy. A culture of trust wherein open communication and dialogue contribute to individual and collective learning is important. Work engagement is conducive to proactivity: it has the potential to broaden employees' horizons, making the workforce more prone to explore alternative and innovative paths.[41]

The ideal is to achieve a situation whereby proactive processes of creation, anticipation and individual and organizational learning prevent all work-related accidents and diseases. Then, incidents are no longer necessary for learning to prevent new incidents. To approach this ideal requires room for out-of-the-box thinking and higher-order learning.

For leaders, it is a challenge to show exemplary proactive behaviour and the associated learning attitude themselves, but also to enable the conditions wherein others (those who are led, direct colleagues, superiors and external stakeholders) are invigorated towards proactive behaviour. Unfortunately, proactive behaviour is easier discouraged than stimulated. Traditional management tools such as detailed procedures or promising financial rewards for realizing performance goals tend to undermine intrinsic motivation and proactive behaviour.

For managers, it can be challenging to deal with workers who behave proactively. Managers like to be in control, but proactivity requires a kind of loose control whereby workers are empowered and encouraged to adapt plans or ongoing operations when necessary. They may come up with unexpected ideas or 'out-of-the-box' suggestions for improvement.

Proactivity is usually associated with decision-making and behaviour of people and policies of organizations. But Information Technology (IT) may support proactivity. Through the use of sensors and big data, it is increasingly possible to recognize patterns and the development of scenarios relevant for early SHW. This implies that IT can generate information that supports anticipation of potential SHW problems in the early stages, contributing to proactivity. The use of artificial intelligence seems to have the potential to suggest new types of solutions, implying new opportunities for improvement.[42] Of course, people remain important for decision-making when such potentials are communicated via information technology.

4.6 MENTAL MODELS IN TIMES OF CHANGE

Vision Zero implies a long-term ambition that clearly goes beyond compliance and beyond the present state of the art. It is known that when you continue to do the things you already did, you will achieve similar results as before, not significant improvements. The same is true for our thinking. When we continue to think about safety and health in the way we have done for decades, we cannot expect to realize much progress. What exactly are mental models? Let's look at an example in the area of flight safety.

LIFE JACKETS IN COMMERCIAL PASSENGER AIRCRAFT ARE JUST AS USEFUL AS PARACHUTES ON FERRY BOATS

Tjabe Smid[43]

When Blériot crossed the Channel in 1909 and when Lindbergh crossed the Atlantic in 1927, it clearly made sense to have life jackets on board. But since World War II, life jackets have not saved any lives in cases of commercial passenger aircraft crashes – even in the case of the well-known emergency landing of US Airways flight 1549 on the Hudson River in 2009. Although worn by some passengers, life jackets were not necessary. But till today, every passenger aircraft is equipped with life jackets, and every passenger receives instructions on how to use them at the beginning of each flight. This is mandatory due to international standards.

The example of the life jackets in commercial passenger aircraft shows how an initially very useful measure or approach becomes generally accepted, can become the standard and continues to be the standard even when it is no longer appropriate. In the meantime, it is engrained in our thinking as a necessity which is more or less automatically accepted. It forms a dominant mental model that influences how (work) places are designed and equipped, and how policies, laws and management systems are developed. In this book, the term mental models is used to address ways of thinking that have often materialized beyond our brains in technical systems, practices, (professional) standards and legislation. A change in mental model therefore implies more than merely a change in mindset.

According to Peter Senge,[44] mental models determine our systems which in turn determine the patterns of events therein. When we want to improve SHW significantly, it is important to change the (production) systems that generate work-related accidents and diseases regularly. If we want to successfully change the systems (institutions, organizations, etc.), we have to re-evaluate our mental models. When seriously considered, it is not hard to recognize that many aspects of the thinking that was developed (successfully!) in the past still persist as important or even dominant mental models and co-determine OSH policies, systems, cultures and performance. This explains quite a few of the SHW problems and challenges we face today.

Mental models play a role in goal determination as well as in problem analysis and choice of approach. Mental models are often a reason to remain doing the things that we used to do, instead of searching for better alternatives. Sometimes it may trigger to *do the wrong things*. For Vision Zero, we have to reflect on our implicit assumptions, leave our comfort zones and silos and develop new mental models that are in line with Vision Zero.

4.6.1 The Roots of Our SHW Mental Models

Many mental models originate from the era wherein occupational safety and health emerged as an organized activity: the second half of the nineteenth century (occupational safety) and the first half of the twentieth century (occupational health). An era of industrialization wherein the realities at work were completely different from those today. Several of the mental models then developed limit the possibilities for SHW improvement. Below is a (very generic) sketch of the context wherein many relevant mental models were developed.

THE CONTEXT WHEREIN MANY MENTAL MODELS FOR SHW WERE DEVELOPED

In the second half of the nineteenth century, most people worked in the new industries or agriculture. Businesses operated locally, producing for local or regional customers and employing local people. Although there were some notable exceptions, most employers were only interested in maximizing profit, not in their people. Safety and health were not regarded as the responsibility of employers.

The new industries were thriving and stood for economic progress. Safety and health risks were accepted as a by-product of production activities. Only when the realities of severe risks could no longer be denied, improvements were made. Industrial jobs were dangerous, dirty and demeaning (now known as 3D jobs) and low paid, physically exhausting and wearing people out. Young, strong and healthy workers were preferred by employers. For most people, life was hard and short due to poor working and living conditions. Most work was unskilled; workers were usually not educated; workers were poor, powerless and often undisciplined. Work did not require much knowledge; tacit knowledge of workers was regarded as non-existing or irrelevant. Many hazards stemmed from dangerous production technologies. Acutely dangerous situations at work were manifold, and serious accidents occurred quite frequently. There was little or no attention to prevention. In case of repetitive serious accidents, some additional measures were taken (e.g., a shield protecting workers from moving machine parts), implying additional costs for the employer. Attention for long-term health impacts was not an issue. For the employer, it was easier and cheaper to replace a worker suffering from an illness or infirmity than to support cure and return to work. Life was short anyway. When work was obviously dangerous, burdensome or stressful, workers received somewhat higher wages or hazard pay to make it acceptable (instead of improving the quality of work).

As a response to the rapid industrialization and poor working conditions, some basic legislation was developed to protect workers, and labour inspections were founded. Legislative requirements were still very limited. Their focus was on acute impacts. Reference to legal requirements was the strongest argument to convince employers to invest in improvements.

In the early twentieth century, *scientific management* emphasized the benefits of a strong division of labour (specialization). Safety engineers were appointed in hazardous industries. Most responsibilities for safety were delegated to these safety engineers, allowing the employers to focus on their role as entrepreneurs. Later occupational physicians got responsibilities for occupational health.

In the context of the nineteenth century and the beginning of the twentieth century, the mental models developed were useful, powerful and innovative, but – similar to the life jackets in airplanes – they still influence our thinking and acting. Mental models are internalized and tend to confirm themselves and persist over time. For significant improvements, we need new mental models that open up new perspectives and opportunities. Examples of mental models that were useful in the past but no longer contribute to solving SHW problems are given in Table 4.3.

The examples above illustrate the relevance to reconsider how our mental models govern our practices, standards, legislation, ideas and habits in SHW.

TABLE 4.3
Examples of mental models that reflect the past but co-determine the present

Successful thinking of the past	May contribute to problems in the present
Safety and health risks are regarded byproducts of production activities.	SHW is not a top priority for (top) managers. Business decisions are often taken without considering SHW. Paying attention to SHW is not considered good for a business career. SHW is regarded mainly a workers' interest. SHW issues are regarded as problems to be solved, more or less apart from the production arrangements; this often triggers a 'let's fix it' mentality. SHW professionals usually focus on 'the sharp end' and do not see top management as their primary target group.
Work serves primarily to earn money.	People tend to accept SHW problems as part of their job. For many young people, having a job and earning money is more important than good SHW. Higher wages or the prospect of a career convince people to accept unfavourable working conditions.
Most work was unskilled; workers were usually not educated and undisciplined.	People are seen as part of the (SHW) problems, not as part of the solutions. Workers are given little control over their work processes. Human error is seen as the main challenge for safety. Safety seems to require 'behavioural control' of the workers (rules, sanctions or behaviour-based safety programmes). Tacit knowledge of the workers is often undervalued and underused. Participation is frequently ignored or regarded as useless or expensive.

TABLE 4.3 (Continued)
Examples of mental models that reflect the past but co-determine the present

Successful thinking of the past	May contribute to problems in the present
Many hazards stemmed from dangerous production technologies.	SHW activities focus on reducing existing risks, not on preventing that they arise. Safety is still regarded as more important than health and wellbeing (though in reality that is often different). The focus on technological hazards often implies having a blind spot for human behaviour and psychosocial hazards.
In existing production systems, additional measures were preferred.	Employers still prefer administrative controls and PPEs over more effective prevention measures. The preference for added-on measures explains why prevention is often regarded as a cost factor.
For the employer, it was easier and cheaper to replace a worker suffering from an illness or infirmity than to support cure and return to work.	Many costs of poor SHW are still shifted to workers, their families and society at large.
Reference to legal requirements was the strongest argument to convince employers to invest in SHW improvements.	Many companies still focus on compliance. This implies a culture of external motivation (you have to) instead of internal motivation (we want to). Compliance is often more important than good SHW. Many stakeholders, including most trade unions, still see more legislation and enforcement as the best ways to improve SHW. Many SHW professionals prefer to refer to legislative requirements rather than to sound business aspects of good SHW.
'Scientific management' emphasized the benefits of a strong division of labour (specialization).	Many talents of people remain unused or underused because they seem not relevant for their job. SHW is organized in silos; these silos often do not communicate and cooperate optimally. The presence of SHW experts is often used as an excuse not to invest in SHW expertise in the line organization. Work organization is often ignored as an important determinant of SHW. Problems at system level are difficult to address and solve.

4.7 LEARNING

Learning is important for improvement and innovation. Learning can be formally organized, i.e., in the form of education and training. It can also take place informally, through processes of learning by doing. Learning takes place not only in individuals but also at the team and organizational levels. Collective learning is, however, more than the sum of individual learnings.

Learning processes can be relatively limited or more in-depth. The distinction between adaptive and innovative learning is hereby relevant (see Figure 4.4). The figure shows the importance of feedback and closing the learning cycle for learning processes. Adaptive learning and innovative learning are both useful. Adaptive learning is useful for routine issues and optimization. Innovative learning is useful for more fundamental, often more complex problems and for innovation. When a problem is detected and corrected without questioning or altering the underlying values or assumptions of the system, it concerns adaptive learning. In innovative learning processes, the underlying values or mental models are also considered. A difficulty for innovative learning is that it implies *letting go* of one or several existing assumptions, values or mental models. Unlearning is much more difficult than learning something new.

The model clarifies the importance of feedback for learning and improvement, thereby closing the learning loop. It also clarifies the importance of reflection on the (implicit) assumptions of the mental model and the values for innovative learning.

Learning by experience or learning on the job is relevant for personal growth and development, both for individuals and for teams and organizations. Kolb described the process of (individual) learning by experience as a four stage cyclical process of

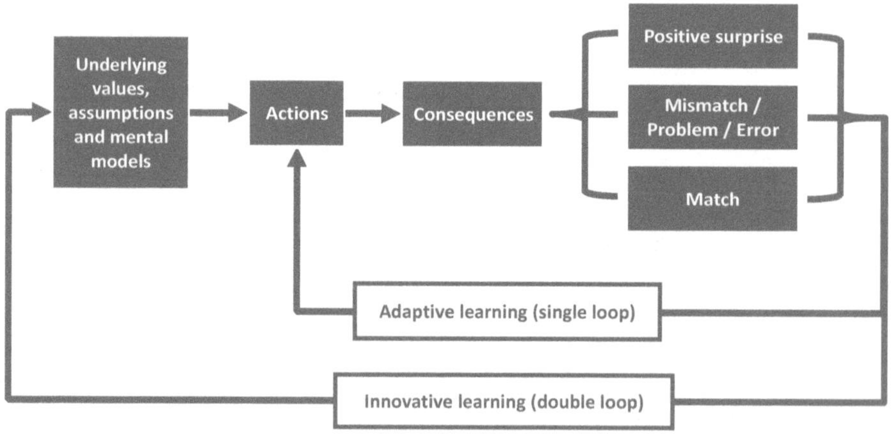

FIGURE 4.4 Adaptive and innovative learning. (Based on Argyris, Chr., (1992) *On Organizational Learning*. Oxford: Blackwell Business; Edgar Schein and Peter Senge call this respectively adaptive and generative learning, see: Senge, P.M., Schein, E., de Geus, A., Gallway, T., (1998) *The New Workplace – Transforming the Character and Culture of Our Organizations*. Waltham, MA: Pegasus Publications.)

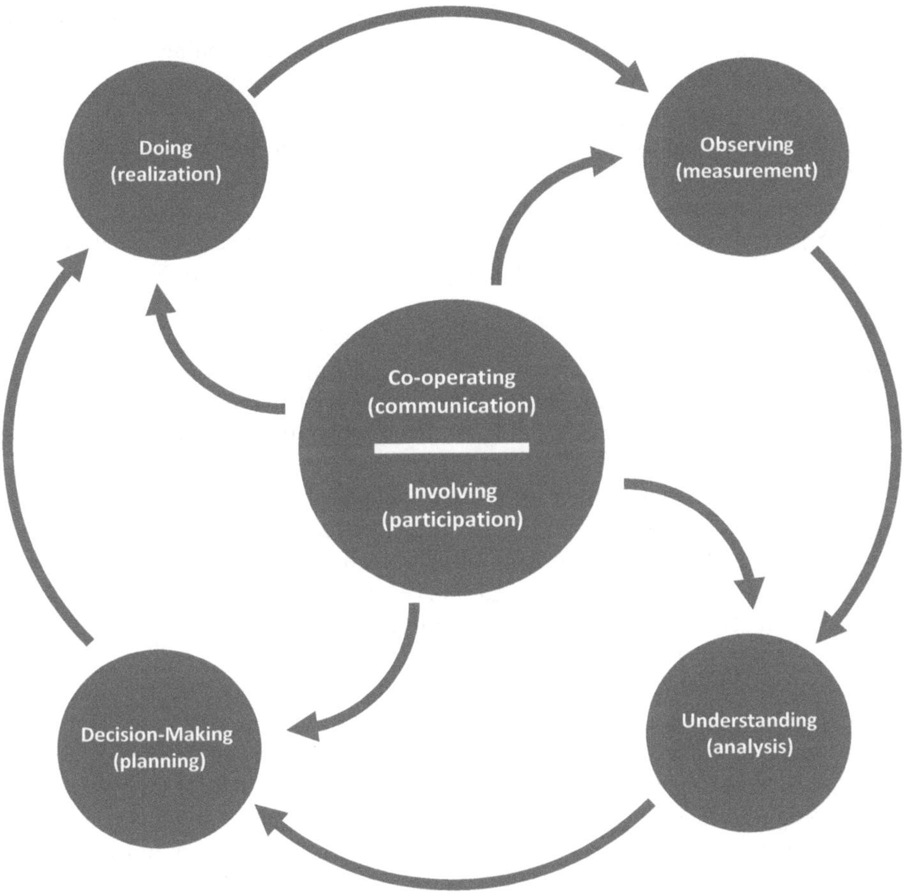

FIGURE 4.5 The model for organizational learning from experience. (From Zwetsloot, G.I.J.M., (1995) Improving cleaner production by integration into the management of quality, environment and working conditions, *Journal of Cleaner Production*, 3(1–2), 61–66.)

observing, analysing, decision-making and doing.[45] The learning cycle has much in common with the Deming cycle of plan, do, check and act. For collective learning by experience (in teams or organizations), cooperation and communication among those involved is also important (see Figure 4.5).

The model in Figure 4.5 is also useful to understand the limitations of a strong division of labour. In the extreme form, it may result in managers taking the decisions and making plans, workers doing the work to realize the plans, supervisors doing the observations and (SHW) staff analysing to understand SHW issues. Then, there are four shifts of responsibility to close the learning cycle, which are four possible hurdles for the collective learning process. It implies that individuals are involved only in one of the four stages of experiential learning that impedes individual growth and development. The model shows the importance of participation of workers (and sometimes external stakeholders) for the collective learning process. Involving the workers implies that their experiences, observations and understandings can contribute to the quality of decision-making and planning. When these are communicated via dialogues, it may also trigger reflections on implicit assumptions and values – a characteristic of innovative learning. In many of today's organizations, a strong division of work has led to over-specialization and the existence of silos between departments and disciplines, impeding (higher-order) organizational learning processes. Interfaces between jobs, disciplines and departments often cause problems for SHW, quality and productivity. For Vision Zero, this implies serious attention to the (re)design of the work organization.

To unlearn old habits that are no longer regarded as acceptable or relevant or to adopt new mental models requires innovative learning. That requires more than merely education and training and can best be achieved through participation in decision-making and meaningful dialogues. Indeed, that implies more open collective learning processes.

Team and organizational learning are enhanced by the social capital in organizations. Social capital is important for collective learning, for managing health and wellbeing and for the functioning of organizations. It refers to reciprocity and trust in relations among work mates and supervisors and requires active participation of the individuals.[46] See Table 4.4 for questions that may support reflection on organizational learning and knowledge management.

It is relatively easy to organize education and training focusing on individual learning, especially for increasing knowledge and adaptive learning. But often training and education have only a limited impact. It is important that the knowledge that is presented can be applied in regular work activities. The team culture may have a stronger influence on the learner than the training. It is therefore important that other team members are equally involved in the training. The opportunities for *learning by doing* are greatly influenced by the organization of work.

TABLE 4.4
Questions for reflection about organizational learning processes

- Is our SHW knowledge regularly updated?
- Are the SHW knowledge and associated learning processes managed consciously?
- Who are our knowledge carriers?
- Do we know what we do not know, but should know?
- Do we know where relevant knowledge can be found (internally or externally)?
- When necessary, do we acquire external knowledge?
- Is relevant knowledge being disseminated adequately?
- Is there an organizational memory, also for the mid- and long term?
- Is critical but positive thinking appreciated?
- Do we involve the workers in key decisions about work and work processes?
- Does everybody in our organization receive the information and resources to work safe and healthy?
- Is it possible to discuss feelings and emotions?
- Is there sufficient social support when people address sensitive issues?

An interesting topic is how organizations can learn from the tacit knowledge of workers.[47] Several methods exist to use the knowledge and experience of workers to improve processes in the organization. Quality Circles and Kai Zen[48] are probably the most well known, and often also address SHW issues. Health promotion circles are a variation thereof specifically aiming at health and wellbeing at work.[49] See also Section 9.5. on using the practical wisdom of the workers.

Argyris distinguished two different ways to organize collective learning processes, respectively focussing on adaptive and innovative learning.[50] Most organizations organize their learning processes according to the adaptive learning model. A key characteristic is that the goals are defined by management, and everything is done to achieve these goals. In this way, feedback loops confirm the organizational assumptions and behavioural routines but that may inhibit the detection and correction of more fundamental problems. Contradiction or critical questions are not appreciated. To motivate people and reduce resistance, the strategy is to maximize the number of people who experience *winning* and to minimize the number who experience *losing*. The focus is thereby on rational argumentation; emotions and feelings are considered irrelevant. Information technology is frequently used for strengthening centralization and top-down supervision. In such a way technology strengthens adaptive learning but may impede higher-order learning, see Table 4.5.

Organizing learning for innovative learning implies to maximize valid information for all parties concerned, to have free and informed choice and high internal commitment to learning. It implies the possibility to address organizational assumptions, underlying values and the mental models of the various stakeholders. Open dialogues and mutual respect should be included in the process.

Most organizations *say* they organize their learning processes following the model for innovative learning. But the reality is often different from what is espoused. Most organizations that say to organizing learning for innovation are actually predominantly operating in a way that results in adaptive learning. Of course, the difference between organizing according to the two ways of organizing is not a matter of black and white. There are greys in between, but then again, the result will predominantly be adaptive learning. Learning from *what goes right* (Hollnagel, Safety-II[51]) implies adaptive

TABLE 4.5
The characteristics of organizing for adaptive and innovative learning

Organizing Adaptive Learning	Organizing Innovative Learning
Define goals and try to achieve them (unilaterally).	Maximize valid information.
Maximize winning and minimize losing.	Have free and informed choice for all concerned.
Minimize expressing or generating negative feelings.	Have high internal commitment by various stakeholders.
Be rational and minimize emotionality.	Besides ratio, there is a place for feelings and emotions.

Source: Argyris, Chr., (1992) *On Organizational Learning*. Oxford: Blackwell Business.

learning. Innovation or learning to *do the right thing* requires higher-order learning. Learning from what went wrong sometimes triggers higher-order learning. In the development of science, new paradigms (mental models) emerge when the existing ways of thinking are no longer able to solve one or several pressing problems.[52]

Many organizations, also learning organizations, often learn in certain areas, but not in others. Companies can learn and improve in safety while they seem not to learn how to deal effectively with musculo-skeletal problems or with stress and well-being challenges or vice versa. Acknowledging common values and interests, open communication across silos and fruitful cooperation seem the key to overcome such limitations. Interestingly, it is not only possible to learn from the past but also from the future when it emerges.[53] Foresight studies can clarify relevant scenarios and help to be better prepared for upcoming challenges.

4.8 CONCLUSIONS

For a strong prevention culture, it is important that prevention is integrated into the *values, the mindsets and actions* of managers and employees. Values serve as a compass for strategic decisions. Besides SHW as values, there are seven other human and social values that support SHW: Interconnectedness, participation, trust, justice, responsibility, growth and development and resilience.

Aligning business related and social values requires dialogues of management and (representatives of) the workforce. An open attitude is thereby a necessity. Values are multi-layered, like organizational culture. It must be clear what the values mean in practical situations. The values have to be operationalized for the specific organizational context. Sharing values often makes the organization an *employer of choice*.

For SHW, it is most desirable when all managers and workers are self-motivated to contribute to excellent SHW. Intrinsic motivation means that people are motivated for the inherent satisfaction of doing the activity itself. Motivation through identification or integration also comes mostly from the self. A relatively high degree of autonomy, competence and relatedness contributes strongly to internal motivation, i.e., a renewable source of mental energy and work engagement.

Vision Zero implies that *doing the right things right from the start* becomes a leading quality principle. It is thereby necessary to distinguish between *doing things right* and *doing the right things*. While d*oing things right* is a rational and practical principle, *doing the right things* depends on values and principles and has strategic impact.

For Vision Zero, it is crucial to be proactive and forward-looking. Proactivity refers to making things happen, anticipating and preventing problems, or seizing opportunities for improvement. The organization of work is important for proactive work behaviour. Jobs that offer opportunities for controlling the work process create conditions under which individuals are intrinsically motivated to be proactive in their work. Proactivity is a self-initiated effort; ownership is crucial.

It is important to reconsider the mental models we use, as they often reflect successes of the past, not for the future. Mental models have often materialized beyond our brains in systems, practices, standards and legislation and co-determine how these are designed and developed. For Vision Zero, we have to reflect on our implicit assumptions, leave our comfort zones and silos and develop new mental models that are in line with the world of change.

Individual and collective learning is important for improvement and innovation. Learning can be formally organized, i.e., through *education* and *training*. There is also the informal process of *learning by doing*. Most organizations organize their learning processes in a way that the fit generates adaptive learning and not innovative learning. It is then important to increase the organizational capacity for learning. In fact, several aspects of Vision Zero described in Chapter 3 (e.g., synergies) and in the former paragraphs of this chapter require higher-order learning processes. That is also the case for several topics addressed in the next chapters.

NOTES

1 Zwetsloot, G.I.J.M., van Scheppingen, A.R., Bos, E.H., Dijkman, A., Starren, A., (2013) The core values that support health, safety and well-being at work, *Safety & Health at Work*, 4, 187–196. Doi: 10.1016/j.shaw/2013.10.001.
2 Beer, M., (2009) *High Commitment, High Performance: How to Build a Resilient Organization for Sustained Advantage*. San Francisco, CA: Jossey-Bass.
3 Beer, M., (2009) *High Commitment, High Performance: How to Build a Resilient Organization for Sustained Advantage*. San Francisco, CA: Jossey-Bass.
4 Zwetsloot, G.I.J.M., van Scheppingen, A.R., Bos, E.H., Dijkman, A., Starren, A., (2013) The core values that support health, safety and well-being at work, *Safety & Health at Work*, 4, 187–196. Doi: / 10.1016/j.shaw/2013.10.001.
5 Gerstein, M.S., (2008) *Flirting with Disaster – Why Accidents Are Rarely Accidental*. New York, London: Union Square Press.
6 Reason, J., (1997) *Managing the Risks of Organizational Accidents*. Farnham: Ashgate.
7 Dekker, S. W., (2017) *Just Culture*. Boca Raton: CRC Press.
8 Schelvis, R.M.C., Zwetsloot, G.I.J.M., Bos, E.H., Wiezer, N.M., (2014) Exploring teacher and school resilience as a new perspective to solve persistent problems in the educational sector – A case of the Netherlands, *Teachers and Teaching, Theory and Practice*, 20 (5), 622–637. Doi: 10.1080/ 13540602.2014.937962.
9 Weick, K.E., Sutcliffe, K.M., (2007) *Managing the Unexpected: Resilient Performance in an Age of Uncertainty* (2nd ed.). San Francisco, CA: Wiley.
10 Beer, M., (2009) *High Commitment, High Performance – How to Build a Resilient Organization for Sustained Advantage*. San Francisco: Jossey–Bass.
11 Brown, J., (2005) *The World Café – Shaping Our Futures through Conversations That Matter*. San Francisco: Berret-Koehler Publishers.
12 Senge, P., Scharmer, C.O., Jaworski, J., Flowers, B.S., (2004) *Presence – Human Purpose and the Field of the Future*. Cambridge, MA: Society for Organizational Learning.
13 Milliman, J., Czaplewski, A.J., Ferguson, J., (2003) Workplace spirituality and employee work attitudes – An exploratory empirical assessment. *Journal of Organizational Change Management* 16 (4), 426–447. Doi: 10.1108/09534810310484172.
14 De Geus, A., (1998) The living company: Extending the corporate lifeline, in: Senge, P.M., Schein, E., de Geus, A., Gallway, T., (eds), *The New Workplace – Transforming the Character and Culture of Our Organizations*. Waltham, MA: Pegasus Publications. pp. 91–111.
15 Ryan, R.M., Deci, E.L., (2000) Self-determination theory and the facilitation of intrinsic motivation, social development, and wellbeing. *American Psychologist*, 55 (1), 68–78. Doi: 10.1037/ 0003-066X.55.1.68.
16 De van der Schueren, M., Zwetsloot, G.I.J.M., (2016) Drie redenen waarom mensen zich zonder betaling ten volle inzetten (Three reasons why people commit themselves fully without any payment),

Tijdschrift voor Toegepaste Arbowetenschappen (Journal of Applied Occupational Sciences), 29 (2), 55–60.

17 Ryan, R.M., Deci, E.L., (2000) Self-determination theory and the facilitation of intrinsic motivation, social development, and wellbeing. *American Psychologist*, 55 (1), 68–78. Doi: 10.1037/0003-066X.55.1.68.

18 Ryan, R.M., Deci, E.L., (2000) Self-determination theory and the facilitation of intrinsic motivation, social development, and wellbeing. *American Psychologist*, 55 (1), 68–78. Doi: 10.1037/0003-066X.55.1.68.

19 Pink, D.H., (2009) *Drive – The Surprising Truth about What Motivates Us*. New York: Riverhead books.

20 Parker, D., Lawrie, M., Hudson, P., (2006) A framework for understanding safety culture. *Safety Science*, 44, 551–562. Doi: 10.1016/j.ssci.2005.10.004; Hudson, P., (2007) Implementing a safety culture in a major multinational, *Safety Science*, 45 (6), 697–722. Doi: 10.1016/j.ssci.2007.04.005.

21 Graeber, D., (2018) *Bullshit Jobs: A Theory*. New York: Simon and Schuster.

22 Ryan, R.M., Deci, E.L., (2000) The self-determination theory and the facilitation of intrinsic motivation, social development and well-being, *American Psychologist*, 55 (1), 68–78. Doi: 10.1037/0003-066X.55.1.68.

23 Bakker, A.B., Dermouti, E., Sanz-Vergel, A., (2023) Job demands and resources theory: Ten years later. *Annual Review of Organizational Psychology and Organizational Behaviour*, 10, 25–53. Doi: 10.1146/annurev-orgpsych-120920-053933.

24 Zwetsloot, G.I.J.M., de Bruin, T., (2022) Surviving relatives as stakeholders for corporate social responsibility and as leaders for meaningful safety improvement – A case study from the Netherlands, *Safety Science*, 157, 105927. Doi: 10.1016/j.ssci.2022.105927.

25 Ryan, R.M., Deci, E.L., (2000) Self-determination theory and the facilitation of intrinsic motivation, social development, and wellbeing. *American Psychologist*, 55 (1), 68–78. Doi: 10.1037/0003-066X.55.1.68.

26 Pink, D.H., (2009) *Drive – The Surprising Truth about What Motivates Us*. New York: Riverhead Books.

27 Leveson, N., (2020) Safety *III – A Systems Approach to Safety and Resilience*. Cambridge: MIT.

28 Flyvbjerg, B., (2022) Heuristics for Master Builders: Fast and Frugal Ways to Become a Better Project Leader. *Saïd Business School Working Paper*, University of Oxford.

29 Juran, J.M., (1964) *Managerial Breakthrough*. New York: McGraw-Hill. p. 139.

30 Conti, T., (1993) *Building Total Quality*. London: Chapman & Hall; Zwetsloot, G.I.J.M., (2003) From management systems to corporate social responsibility, *Journal of Business Ethics*, 44, 201–207. Doi: 10.1023/A:1023303917699.

31 Appicharla, S.K., (2014) Application of cognitive systems engineering approach to railway systems. *Selected Topics in Safety and Development*. Rijeka, Croatia: Intech Open Publishers.

32 Yates, G.R., Hochman, R.F., Sayles, S.M., Stockmeier, C.A., (2004) Sentara Norfolk General Hospital: Accelerating Improvement by Focusing on Building a Culture of Safety. *The Joint Commission Journal on Quality and Safety*, 30 (10), 534–542. Doi: 10.1016/S1549-3741(04)30063-8.

33 Parker, S.K., Bindl, U.K., Strauss, K., (2010) Making things happen: A model of proactive motivation. *Journal of Management*, 36 (4), 827–856. Doi: 10.1177/0149206310363732.

34 Parker, S.K., Bindl, U.K., Strauss, K., (2010) Making things happen: A model of proactive motivation. *Journal of Management*, 36 (4), 827–856. Doi: 10.1177/0149206310363732.

35 Griffin, M.A., Parker, S.K., Mason, C.M., (2010) Leader vision and the development of adaptive and proactive performance: A longitudinal study. *Journal of Applied Psychology*, 95 (1), 174–182. Doi: 10.1037/a0017263.

36 Frese, M., Fay, D., (2001) Personal initiative: An active performance concept for work in the 21st century. *Research in Organizational Behaviour*, 23, 133–187. Doi: 10.1016/S0191-3085(01)23005-6.

37 The most well-known example is the Finnish Vision Zero Forum; see www.ttl.fi/en/services/risks-and-safety-of-the-work-environment/vision-zero-forum-networks-empowering-development (Accessed 9 December 2024).

38 Weick, K.E., Sutcliffe, K., (2007) *Managing the Unexpected*. San Francisco: Jossey-Bass.

39 Parker, D., Lawrie, M., Hudson, P., (2006) A framework for understanding the development of organizational safety culture. *Safety Science*, 44 (6), 551–562. Doi: 10.1016/j.ssci.2005.10.004; Hudson, P., (2007) Implementing a safety culture in a major multinational, *Safety Science*, 45 (6), 697–722. Doi: 10.1016/j.ssci.2007.04.005.

40 Parker, S.K., Bindl, U.K., Strauss, K., (2010) Making things happen: A model of proactive motivation. *Journal of Management*, 36 (4), 827–856. Doi: 10.1177/0149206310363732.

41 Bakker, A.B., Dermouti, E., Sanz-Vergel, A., (2023) Job demands and resources theory: Ten years later. *Annual Review of Organizational Psychology and Organizational Behaviour*, 10, 25–53. Doi: 10.1146/annurev-orgpsych-120920-053933.

42 IEC, (2020) *White Paper: Safety in the Future.* Geneva: International Electrotechnical Commission. https://webstore.iec.ch/en/publication/67876 (accessed 17 December 2024).

43 Tjabe Smid was many years the Occupational Safety Manager of Royal Dutch Airlines KLM. The statements were first formulated in: Smid, T., (1993) *Exposure to Organic Dust and Respiratory Disorders.* Dissertation Wageningen University.

44 Senge, P., (1990) *The Fifth Discipline, The Art and Practice of the Learning Organisation.* New York: Doubleday.

45 Kolb, D.A., (1984) *Experiential Learning – Experience as the Source of Learning and Development.* New Jersey: Prentice Hall.

46 Oksanen, T., Kouvonen, A., Kivimäki, M., et al., (2008) Social capital at work as a predictor of employee health: Multilevel evidence form work units in Finland. *Social Science and Medicine,* 66, 637–649; Van Scheppingen, A., de Vroome, E.E.M., Ten Have, K.C.J.M., Zwetsloot, G.I.J.M., Bos, E., van Mechelen, W., (2013) The associations between perceived organizational social capital, health and employees' functioning in two Dutch companies, *Journal of Occupational and Environmental Medicine*, 55 (4), 371–377. Doi: 10.1097/JOM.0b013e31828acaf2.

47 See Podgorski, D., (2010) The use of tacit knowledge in occupational safety and health management systems. *International Journal of Occupational Safety and Ergonomics*, 16 (3), 283–310. Doi:10.1080/10803548.2010.11076845.

48 Imai, M. (1986) *Kai Zen.* New York: McGraw-Hill.

49 Schoër, A., Sochert, R., (2000) *Health Promotion Circles at the Workplace – Models and Practical Implementation.* Essen: BKK Bundesverband.

50 Argyris, Chr., (1992) *On Organizational Learning.* Oxford: Blackwell Business. Argyris used the term Model I and Model II for the two ways of organizing organizational learning processes.

51 Hollnagel, E., (2014) *Safety I and Safety II. The Past and Future of Safety Management.* Farnham: Ashgate.

52 Kuhn, T.S., (2012) *The Structure of Scientific Revolutions.* Chicago (50th Anniversary edition, original published 1962). Chicago: University of Chicago Press.

53 Senge, P., Scharmer, C.O., Jaworski, J., Flowers, B.S., (2004) *Presence – Human Purpose and the Field of the Future.* Cambridge, MA: the Society for Organizational Learning.

5 Leadership
The Art of Using Power Wisely

Vision Zero requires leadership. Leadership was therefore already mentioned in the two former chapters, and it is mentioned throughout this book as a factor that is interwoven with almost every aspect of Vision Zero. Leadership, as far as relevant for Vision Zero, is addressed in this chapter. We start with a CEO's perspective on leadership and Vision Zero based on an interview with Peter Spit.

5.1 A CEO'S PERSPECTIVE ON LEADERSHIP AND VISION ZERO

Peter Spit is the CEO of Schunk Xycarb Technology (SXT).[1] SXT is a global supplier of Graphite, Ceramics and Quartz-ware solutions for the Microelectronic industry. With production facilities in Europe, Asia and America, SXT operates globally; its headquarter is based in The Netherlands. The company belongs to the German Schunk holding. What is his perspective on Vision Zero?

5.1.1 What Is Your Motivation for 'Zero'?

We cannot allow that some of our processes are not in good order. As a leader, you have a responsibility for the people and their families. Going for zero accidents means your ambitions are clearly expressed and valid for a longer period. You can assess gaps with that ambition. That triggers improvements that are otherwise difficult to achieve. Safety and health must be part of your DNA so that its importance goes without saying and is always there.

I wanted to improve safety when I came into office as CEO. But it was accelerated shortly thereafter by a fatal accident and a fire in 2014. That was a difficult year. It triggered many discussions: How could it happen? Who is accountable for what? It was important to take the time to discuss that. There were emotions we had to deal with. Surviving relatives advocated for more prevention and touched the hearts of our community.[2] It was necessary to 'jump over our own shadow' to rethink our habits and ways of working. That required time for reflection from the team of directors, including those with the prime responsibility for sales or finance.

DOI: 10.1201/9781003621225-5

Looking back, the question can be asked: suppose that no accidents happened in 2014, What would have been different? What would we have improved in that case? Probably we would have made the same development, but slower.

Health and wellbeing are more comprehensive than safety. You can only partly influence these. It is often a matter of long-term exposures and impact, but there are overlaps and interactions with safety. We want to ensure that people can do their work for a long period and stay healthy. In the United States and Taiwan, we work with chemicals like hydrogen fluoride requiring people to wear adequate personal protective equipment. We improve the ergonomics of our workplaces, such as flexibility in the offices for sitting or standing at work. People also have to take responsibility for their own health, for instance, concerning lifestyle issues like nutrition and physical exercise. You can only partly influence that by supporting measures.

5.1.2 What Are Your Main Achievements?

First of all, a cultural change. There is now awareness and openness. The commitment to zero created an obvious platform to discuss minor incidents, errors and surprises. People may err, that is human. But it is vital that it can be freely discussed. People now discuss such things without fear, including people with temporary contracts. Of course, sometimes there is a question: Does what you did conform to our standards? But the intention is always to learn and improve. People are involved; there are discussions and notifications. The number of notifications has increased significantly. People also know they can address each other on safety, which is broadly accepted and practiced.

It takes years to achieve this; it goes very slowly. Sometimes the organization seems to take a step backwards before you can make the next step. The desired behaviour is sometimes also a matter of training and skills, for instance, taking the responsibility when working with contractors.

Secondly, safety is integrated into our innovation and development processes. We check our technologies much better than before. In 2018, we changed to a new coatings process. The old process was at 1,400 degrees Celsius, and the new process was at 2,200 degrees. Then you have to be sure that the reactors and other equipment are suitable for that temperature. In case of changes and innovation (in chemistry, machines, automation, etc.), safety and health is always an issue. You can only go over to the next phase in a development process when safety and health are adequately addressed.

5.1.3 What Made a Difference?

The commitment of senior management is essential. The top has to be the SHW champion. That is crucial in getting the full commitment of the middle managers. They have so many other things to work on. It is partly top down, but it is also a matter of orchestrating. We were inspired by a meeting at Urenco, the uranium enrichment facility in the Netherlands. They practiced always *Safety First*, in each meeting. That requires discipline and perseverance. It is important to realize that there are always things that can be improved. Safety and health are never ready. Again and again, people see and notify things that you did not know that they existed.

What also made a difference were trainings where people were able to experience for themselves how it is to lose their thump, or to be in a wheelchair. It made them feel how valuable safety and health really are.

People sometimes want to do things (too) quickly. It can raise emotions when that is questioned. Sometimes the opposite is asked: are we not exaggerating about safety and health? In such cases, we exchange visions and discuss potential problems: What if? These discussions are sometimes difficult. There are moments when you do not have a clear answer. It is then important to have a good dialogue and to exchange visions before decisions are taken.

5.1.4 What Is Your Takeaway for Business Leaders?

Take safety and health seriously, always. It needs continuous attention; it is never ready. It needs to be part of your DNA. Always remember: as a CEO you are accountable!

5.2 SENIOR LEADERSHIP AND THE CHALLENGE OF CONTROL

Leadership is often understood as leadership by senior management. Senior leaders have an important role to play in improving SHW, but at the same time they are responsible for the continuity of the organization and its financial performance. They are responsible for the people employed, the status of their organization in society and so on. The challenges they face are much broader than merely leading SHW. They are also responsible for ensuring compliance with legal obligations and organization principles and guidelines. They must ensure the organization's *license to operate* which is heavily dependent on the confidence of external stakeholders, especially authorities. Business continuity and good financial performance are key issues (also for safeguarding employment). Authorities increasingly expect organizations – especially high-hazard industries – to be able to demonstrate that (external and internal) SHW risks are always under control, which is easier said or suggested than realized. Customers increasingly see good SHW performance as an indication of good management, which they consider important for the quality of the product they expect or the reputation of the organization. Authorities and external stakeholders like governmental organizations and NGOs increasingly expect organizations to also ensure that in their value chains SHW and associated labour rights meet international standards. The challenge to achieve zero accidents or work-related ill health is appealing to senior managers, because it fits very well in the idea of being in control. But it requires hard work, genuine personal commitment and perseverance to achieve it. Interestingly, it does not require strong top-down control, but some form of loose control, two-way communication and the commitment of the workforce.

To the outside world, senior managers are the personal face of the organization. The confidence in the organization of authorities, of business partners and of society at large is greatly influenced by these leaders. Senior managers are expected to be 'in control' of their organization. Apart from initiating some changes, an important aspect of their job is to be and stay in control and to communicate externally as well as internally that they are in control.

EXTERNAL COMMUNICATION BY SENIOR LEADERS

In case of emergencies or serious events, senior leaders will be hunted by the press. Many senior managers receive media training: to be able to communicate in such conditions that they are still in control. The first strategy is to downsize the problems. They then communicate that the problems occur only incidentally and that these imply some (minor) challenges. For senior leaders, it is important to avoid bad publicity for the organization, the brand or themselves. To be out of control is taboo for the external stakeholders. Internally, SHW problems need to be discussed and addressed adequately.

Internally senior leaders often prefer that problems and solutions are summarized on one A4 page. Most senior leaders will communicate sensitive issues only with a few confidants.

For an organization controlling the processes to generate the products or services for their clients is vital. However, too much focus on control is at odds with opportunities for innovation, experimentation and learning, and the need to regularly adapt the organization to new external circumstances. This balancing act between delivering and change also implies getting people to change, probably the most challenging task for any manager.

Traditionally, *being in control* presupposes that things can be assessed objectively and clear cause-and-effect relationships determine the impact of actions. Unfortunately, today's world is not compatible with that ideal. Often external influences come unexpectedly and require instant response. The introduction of new technologies or new ways of working may unexpectedly introduce new and unknown risks. Human behaviour is never fully predictable. In case of conflicts, there are usually two sides of the coin and two opposed perceptions, not one objective reality. Then it makes no sense for leaders to pretend that they are Mr or Mrs Knowitall, who knows what needs to be done. Especially in case of uncertainties, ambiguities and complexities, it is important to involve others, verify different perceptions and take decisions only after dialogues with the most important stakeholders.

5.3 THE IMPORTANCE OF LEADERSHIP

A lot of research has focused on leadership styles, demonstrating that transformational leadership is most effective. Transformational leadership refers to the leader moving the follower beyond immediate self-interests through idealized influence, inspiration, intellectual stimulation and individualized consideration.[3] Vision Zero is important for idealized influence, inspiration and intellectual challenges. Nevertheless, knowledge of leadership in safety or SHW is still rather limited.[4]

Leaders should adopt Vision Zero and the ambition to prevent all (serious) accidents and ill health and to promote excellence in SHW. This cannot be a paper exercise only. Senior leaders should actively contribute to the development of plans and participate in activities to improve SHW prevention.

There are four strategic reasons why leadership is important for Vision Zero:

- Going beyond compliance requires giving direction to the development of the organization and the perseverance to keep that direction, supporting the development of a prevention culture (see the CEO experience in Section 5.1 and Chapter 7), promoting proactivity and creating the conditions for organizational learning.
- Direction is important for Vision Zero as it presupposes that the organization employs a continual process of strengthening prevention and promotion in the area of SHW, thereby reducing accidents and work-related ill health to approach or achieve the zero outcome. Leadership by senior leaders, but in fact at any organizational level, is essential for a prevention culture and for supporting proactivity and organizational learning.
- Middle managers and first-line supervisors have an important role to play in managing SHW as well. They are the linking pin between the higher echelons and the sharp edge where the risks of SHW most prominently appear. Their loyalty usually goes to both sides. That is a difficult position. Middle managers are often overloaded with production-related tasks; nevertheless, it is important to involve them in the development and planning of SHW activities. They know the intentions of higher management, and, at the same time, they are aware of the risks, risk perceptions and habits of the operational staff in the workplace.
- On the shopfloor, there are usually informal leaders. They can be advocates of SHW.[5] Formal leaders may empower them as advocates of SHW, use their intrinsic motivation and give them room and support to act as SHW champions. Specifically, to their peers, informal leaders can play an exemplary role, by being proactive and sharing their motivation for SHW with others on an operational level.

5.4 PERSONAL COMMITMENT

Goal commitment can be enhanced by leaders communicating an inspiring vision and behaving supportively.[6] Visible personal commitment to SHW improvement by senior leaders is crucial to give *direction*. Visible commitment means in practice: showing exemplary behaviour, propagating the value of SHW, involving people and sincerely listening to them and making resources available (time, money, knowledge, tools).

Commitment to Vision Zero and SHW improvement depends on personal values, professional competence and SHW knowledge and experiences with SHW issues. Personal values are influenced by the social and human values that were considered important during childhood. Training and education in cultural values, knowledge and skills relevant to SHW enhance competences and can induce personal conviction about the value of SHW. Personal experience with a serious accident can be a life-changing event. As a result, people may become intrinsically motivated to improve safety for the rest of their lives. Similarly, the personal experience with physical or mental health problems – of themselves or people close by (e.g., family members, friends or close colleagues) – often results in acknowledging the value of SHW and strengthening the intrinsic motivation to improve SHW. The personal commitment expresses itself also in terms of urgency to come into action. Urgency is important when an incident takes place, but also when the organization is warned or sanctioned for non-compliance by the authorities.

Management commitment can be strengthened by organizational policies, programmes to develop SHW leadership and increased understanding of the value case for SHW.[7] For leaders, it is useful to reflect on their personal values and commitments and what that implies for leading others. Openly discussing the ethical aspects of SHW and associated ethical dilemmas can have a very positive impact on the credibility of leadership commitment in the eyes of the workforce. Regular reflection on what went well and what went not so well is another important aspect of leadership: it is vital for the timely adaption of activities and for learning.

Senior leaders work relatively far from the operational level, i.e., far from the sharp end. How then to be the example, to set the tone for the organizational culture through visible personal commitment? How to be a good role model? There are not many occasions to show personal commitment in a convincing way to operators, i.e., not only in words. One such occasion is during site visits and workplace visits. These are highly valuable opportunities to communicate directly with people working at the edge. It is recommended to perform such site visits regularly and frequently, say 15 to 20 times per year. Instead of telling people what to do, it is crucial to create an open conversation between senior management and operational staff about their work the problems they encounter, the opportunities they see, and what is important in their jobs. Asking open questions to get a better understanding of the challenges operational people face in their work is important. Meaningful dialogue triggers mutual learning and is a great opportunity to value the practical wisdom of operational workers and to show genuine commitment to SHW. Of course, in the direct two-way communication with people on the shopfloor, senior leaders will also answer questions from operatives and refer to the organization's values and policies and their personal and organizational commitments. Showing exemplary SHW behaviour communicates strongly that SHW is not about words and slogans but about actual behaviour and personal commitment, see Table 5.1.

Unfortunately, some senior managers still regard workplace visits as *workplace inspections* whereby the leader should check the behaviour of the worker and correct it if deemed necessary. The unintended message to the worker may be: your opinion and experience are less relevant than my status. That impedes good communication on SHW matters.

Leaders should realize that their acts, decisions and communications can easily have a symbolic meaning for the workers as if they are under a magnifying glass. This

TABLE 5.1
Some characteristics of exemplary behaviour

- Be consistent in showing SHW-promoting behaviour
- Address personally SHW issues whenever that is appropriate, even when it is against the short-term interest of the organization.
- Always treat people in a fair way and show respect for other people and their experiences at all times.
- Make sure to be visible on the shop floor regularly and take time for meaningful dialogue with operators. Spend the majority of time listening not speaking or sending messages. The aim of listening should be to understand, not to reply.
- Initiate meaningful conversations and dialogues and be a good listener. It is more important to let people think for themselves than tell them what they have to do.
- Show the will to learn from SHW problems and errors, welcome *bad news.*
- Reflect openly about how to deal positively with errors, including those of yourself.

can be positive (see the following story of the paper mill) or negative (e.g., when a manager ignores a safety rule, or when somebody makes career who is clearly not an advocate of SHW). The symbolic meaning of the leader's behaviour can have much more impact than the literal meaning of their words.

THE SYMBOLIC MEANING OF COMMUNICATION: THE STORY OF THE PAPER MILL[8]

'A nice example of genuine personal commitment with long-term impact was demonstrated more than a decade ago in a paper mill in the Netherlands. A serious accident had occurred, wherein a machine of around €50M was involved. The labour inspection came to investigate the accident and suspended the work. After two days of intensive investigation the labour inspectors said something like: "we did not find anything wrong with your machine", and they allowed a restart. The CEO then said something like: "Thank you very much gentlemen, but we are not going to restart production before I know personally what has caused the accident!" A few days later the main cause of the accident was found, and the machine was adapted. More than a decade later, the workers still tell this story to explain that in their organization the leader's commitment to zero accidents is genuine'.

The organizational policy also impacts the motivation of leaders: Is genuine commitment to SHW good for their career? Or are other efforts higher valued in this respect? How was the (SHW) situation when the leader came into office? Were the SHW aspects already in good order? Did the leader inherit quite a couple of nasty SHW problems? Are there incentives in the organization to postpone actions to improve SHW?

Legal arrangements may also play a role in triggering commitment. In an increasing number of jurisdictions, there is now legislation that makes senior leaders personally accountable and liable for serious SHW risks. In case of fatal accidents, they can even be accused of manslaughter. Of course, most organizations have insurance to cover the personal liability of their senior managers. But insurances have their exclusions, e.g., in case the leader was already aware of the risks but did not enough to reduce or control them, or neglected warning signals. Unfortunately, legal arrangements trigger only the external motivation to comply with legislation; legal obligations are not able to trigger the commitment to go beyond compliance which is essential for Vision Zero.

Both senior leaders and lower-ranked leaders have power. The power to overrule people. Having power may trigger the idea that power should be used. But by using power you deprive others of (part of their) autonomy. It may lead to ignoring the values and interests of others which may affect them personally, often emotionally. That easily undermines their motivation and commitment. It may trigger a culture of fear, anxiety and resistance. It is therefore a challenge for leaders to use their powers wisely and humbly with emotional intelligence.[9]

5.5 THE EMPLOYMENT RELATIONSHIP AND MUTUAL COMMITMENTS

5.5.1 Two Visions on Employment Relations

There are two management visions on entrepreneurship and employment relationships, based on different mental models.[10] In the first, traditional vision, an organization is primarily regarded as a production system, more or less detached from its environment, with the aim to earn as much money as possible. In the second vision, the enterprise is seen as a living organism that is part of a broader ecosystem (people, communities, nature, markets, etc.) with a primary goal to contribute to the quality of life. In this vision, making a profit is still very important for the organization to survive but is regarded as a means to contribute to the higher goal.[11] The SHW of the employees and the quality of working life may be part of the purpose of the organization. Shifting social problems to society or the social security system will then be prevented if possible.

Vision Zero is a way to prevent business upsets as well as an expression of care for the employees and their health and wellbeing. It is then logical to see employees also as individuals and to have a good relationship with them. Good personal relationships between the employer and the employee are what we often see in small organizations and family businesses. Many CEOs of family businesses regard the people who work for them as a kind of extended family. They show empathy. They take care of them in good, but also in difficult times. In large organizations with a larger workforce, the employment relationships cannot be that personal. Nevertheless, a large organization can take good care of its employees, and top management can strive for a good relationship with all of its personnel. The personal values of the leaders are hereby of key importance. A CEO may want to be a good employer as part of his or her personal identity. For an organization, good employership means being an attractive employer, because it offers meaningful work and good secondary terms of employment and does not discriminate against certain groups of people from employment. In practice, it means: Respect people; take them seriously. Always keep in mind they are human beings. Recognize that they have first-hand knowledge about the practices and habits in the workplace and the work processes. Do not try to control them or their work unnecessarily. Operational people want to be involved in how their work is organized and want to have a say in the way they should perform their individual tasks. This requires leaders to consider the work organization and organize operator participation in organizing the operations. All too often, work is organized in a way that employees are expected to only execute their work as described by others; that transactional approach is not compatible with Vision Zero.

Any relationship has two sides: two different points of view. Employers expect their employees to be good employees who are loyal and committed to contributing to the goals of the organization. Employees expect their employer to be a good employer and to practice social justice in the employer–employee relationship. That relationship, though unequal in terms of power and interests, can be a good one. But that requires the commitment of leaders to the wellbeing of the employees as well as commitment from the side of the workers to contribute to the organization's objectives.

5.5.2 The Psychological Contract

To better understand the relationship between the employer and the employee, it is useful to consider the so-called psychological contract between management and employees. What do the leaders expect of their operational staff? And what do the employees expect from their leaders? Psychological contracts are mutual agreements about the reciprocal obligations and expectations in employment. The basis is developed during recruitment, but it will evolve due to ongoing, dynamic reciprocal processes during the period of employment. Some parts of the psychological contract can be explicitly negotiated during the phase of recruitment, or in unionized environments be formalized in collective agreements for entire business sectors.

The term 'contract' is a bit confusing here. Usually, there is no formal written contract implying formal obligations. For an employer, there is no formal obligation to meet the expectations and needs of employees, it is merely a matter of ethics and responsibility. It might help to regard the expectations of the employees as the expectations of customers for delivering quality. Indeed, meeting the needs and expectations of internal customers, and taking care of the quality of working life is part of the total quality philosophy.

Rousseau distinguished two types of psychological contracts:[12] those with minimum commitments from both sides and those with higher commitments. The first type of psychological contract is usually found in the case of temporary jobs and/or precarious work and comprises mainly transactional issues (work hours, productivity targets, etc.). In the case of higher-level mutual commitments, the psychological contract also addresses the relationship between management and employees. Mutual

TABLE 5.2
The psychological contract for high-commitment high-performance organizations

Management expectations	Employees expectations
Dedication to mission and strategy	A non-political culture – leaders do the 'right' thing. No hidden agendas
Commitment to organizational goals	Respect Opportunities for personal growth
Loyalty to the organization	Security, good faith efforts to avoid layoffs
Behaviour consistent with organizational values	Leadership behaviour consistent with social values
Unselfish contribution by the employee	Unselfish and non-authoritarian behaviour at the top
High performance and productivity	Meaningful work and social support
Collaboration and teamwork	Co-workers who share common values
Commitment to safety	Commitment to safety, health and wellbeing of the employees
Openness to feedback and learning	Opportunity to speak truth to power and being heard
Initiatives	Room for bottom-up initiatives
Self-management	Participation in goal-setting
Flexibility	Autonomy, social support

Source: Elaborated upon Beer, M., (2009) *High Commitment, High Performance – How to Build a Resilient Organization for Sustained Advantage*. San Francisco: Jossey-Bass.

loyalty and stability[13] are key factors for a high-performance organization;[14] it is also a good basis for Vision Zero.

A number of expectations that can be relevant for mutual commitment are mentioned in Table 5.2.

Research shows that there is often a discrepancy between the respective expectations. Management has the power and opportunities to explicitly express its expectations and to stimulate the employees to meet those expectations. Employees do not have such powers, and sometimes they even apply for a job without fully considering what they could expect from their new employer. For management, it is relatively easy to neglect several expectations of the employees.[15] The result is often a disbalance in meeting mutual expectations, or to put it more strongly: violation of these expectations is often the norm.[16] This generates feelings of being unfairly treated on the side of the employees, which is likely to affect their loyalty to the organization. In more extreme cases, it may lead to an *us versus them* culture wherein there is structural mutual distrust between leadership and workers; that is personally frustrating for all concerned and affects the functioning of the whole organization.

For managing organizations the Vision Zero way, it is important to understand employees' side of the psychological contract very well and to ensure there is sufficient balance between what is expected and delivered from both sides. This requires an open dialogue between management and workers. In that process, the topics mentioned in Table 5.2 can be of assistance. Through open dialogues, mutual understanding and commitment will gradually grow – a sound basis for mutual trust.

The relevance of trust was already mentioned in Section 4.1.1. Trust and the psychological contract combine the two perspectives of leaders and operational staff: trust and distrust are always mutual. Often leaders perceive that the operational staff does not really trust them. Then the question is: Why is that the case? What can leaders do to restore trust? When leaders want to improve trust in the organization, it is vital that the leaders reflect on how far they themselves trust their people. Then it is useful to reflect on the psychological contract and to openly discuss the relevant dimensions with the staff or their representatives. Restoring trust may require a long-term effort anyway, but investing in good mutual relationships pays off, because it also implies an increase in social capital.

Of course, sometimes exceptional situations may arise, wherein decisions have to be taken that contradict the expectations of the employees. In some cases, individual leaders are constrained by organization-wide procedures or pressure from the top. When contradiction with the employees' expectations is a necessity, it is crucial that leaders honestly explain the reasons why and avoids hidden agendas. It is also important to render social support to the employees.

The benefits of a good employment relationship for the organization comprise higher commitment of the employees, greater attractiveness on the labour market, improved organizational culture, better teamwork, higher quality of products and services, higher performance, improved SHW[17] and employees who identify themselves with their work, the organization and its brands. The benefits for the employees comprise a nicer and more open organizational culture, confidence to always be respected and taken seriously, meaningful work and opportunities for personal growth and a good SHW performance.

5.6 PROMOTING AND ENABLING VISION ZERO

For Vision Zero, it is important that leaders and operational staff show proactive behaviours. But that poses a challenge: proactive behaviour, adaptability, creativity and learning cannot be generated by prescription, procedures or by setting targets or rewarding achievements. Proactive behaviour can, however, be enabled by a clear and compelling vision[18] and realizing a supportive organizational context. Next to showing proactive behaviour, leaders should also stimulate adaptability, for instance, by adopting and supporting changes initiated by others.[19] Enabling leadership implies encouraging the dissemination of proactive approaches upward through the formal hierarchical channels but also empowering people and giving them access to resources. Leaders should use their power to create the contextual conditions that are conducive to SHW.

Motivated people are able to make choices themselves and to enjoy a relatively high degree of autonomy. This enables proactive behaviour and also facilitates their engagement. Engaged employees put more effort into their work because they identify with it. They perceive their work as challenging, and they work with a high level of energy and a strong identification with their work. Identification with one's work is conducive to intrinsic motivation. In relation to engagement, it is also important to take away hindrance factors like bureaucracy, role conflicts or other hassles. For individual workers, it is important to know that they are supported by their direct leader and that they will receive social support (e.g., constructive feedback) when needed, besides having opportunities for skill variety and personal growth. Social support should not be limited to individuals and also relate to teamwork and collaboration.

Already in the process of selection, hiring and workplace induction, it is important to ensure that people are motivated to contribute to improvements in SHW. This is even more important when it comes to selecting new leaders. For organizations that adopt Vision Zero, it is vital to have competent and committed leaders. New leaders should be selected on their motivation and track record in SHW. As SHW values are more difficult to change than professional skills and knowledge, verification of the presence of these values is important during the selection process of new leaders.

Even competent leaders may be confronted with ethical and practical dilemmas and unexpected and undesired events. That may include externally triggered surprises (e.g., products not delivered on time) but also consequences of well-intended actions by operational staff. In the latter cases, a 'just culture' is essential, and these negative surprises should be treated with a learning mindset and regarded as opportunities for individual and organizational learning.

The powers of senior leaders should ensure that SHW risks are already addressed in the stages of conceptual decision-making, design and procurement and that problems encountered in the production stage are not only solved but also fed back to those involved in the design stage (see Section 6.5). Leaders are responsible for SHW of all employees and have the (legal as well as ethical) obligation to actively promote improvements thereof. They have to ensure that necessary improvements are implemented timely. They have to be alert to upcoming SHW risks and actively promote organizational learning from SHW-related experiences in their own organization or elsewhere. Finally, it is important to create ways of working and a psychologically safe climate wherein dialogues and discussing opposing statements and ideas are welcomed and where a shared orientation on the future can flourish.

Well-informed decision-making regarding SHW issues requires SHW-specific competence. In most industries, senior managers are qualified in economics, law or business administration. Unfortunately, such studies often lack serious attention to SHW. Consequently, managers must develop these competences during their careers. Organizations should ensure that SHW leadership is an integral part of any leadership development programme. Moreover, it is getting time for business schools and universities to integrate SHW leadership and competence and the importance of a prevention culture into their curricula.

5.7 CHALLENGES FOR LEADERSHIP

5.7.1 Receiving Adequate SHW Information

Senior managers do not work on the sharp edge. For understanding the SHW risks they are accountable for, they cannot rely on their own experiences or first-hand observations. Reporting systems and meetings with line supervisors and operational staff are the main ways to receive relevant information. Receiving adequate SHW information is strongly dependent on the organizational culture and a major challenge for any senior manager. There are many examples where poor reporting about safety critical situations preceded serious accidents. Blaming the reporter or the victim will block the flow of critical information. That allows unsafe or unhealthy situations to persist. The same is true when critical information is not used to take adequate preventive action.

It is difficult to assess the quality of the information that reaches them. In case an anomaly is reported, is it just a symptom of a minor issue, or is it an early warning signal of a (potentially) serious problem that really requires the involvement of senior management? It is of utmost importance that senior managers do not neglect *weak signals* and ensure that proper action is taken by the appropriate line manager/supervisor (see also Section 9.4). Even when they are very busy with other things, it is an essential responsibility to verify information and the associated actions at all times.

Middle managers and line supervisors are also expected not only to be in control but also to report serious problems directly to the senior leaders. But organizational reporting systems tend to confirm that the organization and their leaders are in control. Surveys about the quality of SHW in organizations consistently show that managers are assessing the situation more positively than the workforce. Managers tend to see SHW through a 'rosy' lens. It is not known whether that usually is due to a lack of good information, too much optimism or a form of complacency.

People have to feel free to report activities or conditions that imply an acute danger to safety or health or to handle against the planning and normal ways of working when necessary. It is a challenging task for leaders to create the conditions to make this happen. Communication between (senior) managers and operational people can easily be constrained by unintended hurdles on both sides. As I wrote back in 1994 in my doctoral thesis: *By putting emphasis on systemic problems, managers generally become directly involved in what first was thought to be an operational problem. For managers, it can be rather boring to get involved in such problems, which are perceived as operational. Whereas people who are supposed to solve those operational problems often find it difficult to say to managers 'your behaviour is part of the problem' or 'this problem cannot be solved without your personal involvement'*.[20]

Indeed, when senior managers are confronted with serious or persistent problems in the organization, they are usually 'not amused'. That is why operators may think twice before they report a serious problem. As a result, inconvenient information may be withheld or dismissed. In case of reporting unplanned events or proposals for improvement, it is essential to receive rapid, fair and adequate feedback. It is always better to be taken seriously and to receive disappointing news than getting the feeling the suggestion or report was neglected because it was not welcome. Negative feedback transforms critical thinkers into passive bystanders; Why should people report problems when neither the people who report a problem nor the problems are taken seriously?

Senior management may not be fully aware of critical situations at operational level or expect lower levels in the organization to solve these. For (senior) managers it is therefore crucial to be aware that they and their reporting systems may have *blind spots* regarding SHW problems and opportunities. Compare this with Section 5.1 where the CEO stated that again and again people notify things that you did not know they existed. Critical information and warning signals should be welcomed and taken seriously. Leaders have to be open to unfavourable information, be able to listen carefully to what is said and to what is not said and have a learning attitude. It is also important for managers to have a few people with whom they can openly discuss their doubts and dilemmas in an atmosphere of trust, like the king's jester in ancient times, perhaps a colleague from another organization or an external coach may take such a role.

Kings (in many cultures) were often surrounded by yes-men, who were afraid that contradiction would cost them their lives. The king's jester had the privilege and right to talk and mock freely without being punished. A good sense of humour was thereby indispensable. The jester was the only one allowed to bring bad news or make critical comments. For today's leaders, it is better to regularly invite people to come up with critical information, but humour is still important in bringing such messages.[21]

An effective way to check the information received via the formal reporting systems is to have regular management walkarounds or site visits. They were already mentioned above as opportunities to show commitment. By asking open questions to get a better understanding of the challenges operational people face in their work, this often generates insights that are complementary to those obtained via the formal reporting systems.

5.7.2 Balancing SHW and Productivity

It is a challenge to maintain a good balance between SHW and productivity or profitability. Nowadays many organizations, including commercial companies, have adopted the obligation to include ESG (environmental, social and governance) criteria. SHW is part of the social criteria.

The balance between SHW and productivity varies between concrete situations where priorities have to be set. SHW is always considered important, but the same is true for productivity. Good arrangements of production and work should ensure SHW is part of normal operations. When SHW is ensured, leaders and workers can focus on productivity without compromising SHW. But now and then, at specific moments, planned or unexpected, SHW is no longer self-evident; e.g., when a new technology is introduced, when there are changes in the work organization, when a new employee starts to work or when something happens which is not foreseen. In such cases, SHW deserves special attention. To give priority to SHW when it really matters makes sense from an economic point of view; the costs of prevention are always less than the cost of an accident or ill health, and therefore investing in prevention makes good economic sense.[22]

Organizations should not depend on the choices of individuals to balance SHW and productivity. Senior leaders should give guidelines to supervisors and operational staff and clearly communicate the priority order in case choices have to be made. A Vision Zero organization in the trade sector communicated the following priority order: (1) safety and health, (2) late delivery, (3) quality of products or services and (4) productivity/costs. When asked about the reasons to put safety and health on top, the managers explained that (a) SHW impacts can be irreversible, (b) any problem with SHW also implies extra costs and (c) SHW problems are often associated with delay in delivery as well as quality problems of products and services. It therefore makes sense to have SHW as a top priority. The precise ranking is to be decided by each organization taking into account the SHW risks and the business risks associated with their activities.

5.7.3 Preventing Drift to Danger

Drift to danger is a phenomenon meaning that, in the long run, cost-saving programmes have the tendency to undermine safety margins.[23] When considering cost-saving programmes, it is important to see SHW not as a cost factor but instead as an investment. An investment not only in good SHW but also in good and ethical business. Dealing responsibly with on-going cost-saving programmes to prevent drift to danger is an important challenge for any leader.

Dealing with cost-saving programmes is complex. It requires a strong focus on possible long-term impacts and early warnings of unintended impacts. It requires expertise in all three SHW disciplines and leaders who demonstrate strong SHW leadership.

There are many ways in which cost-saving programmes may undermine safety margins for SHW. Does the programme lead to reducing or postponing maintenance? To replace components or materials with cheaper alternatives? To reduce training programmes? To staff reductions? To an increase of temporary personnel? To (further) intensification of work? To the use of technologies beyond their technical life span? Etc. Many individual cost savings are in itself not problematic, but the combination of several cost-saving measures may lead to serious SHW problems. The joint impact of several cost-saving measures is easily neglected because there is often time pressure and a short-term focus.

Cost-saving programmes may also lead to missing opportunities for improving SHW, e.g., investments in safer technologies, greater expertise in high-risk areas or in a work organization that facilitates learning on the job, personal development and wellbeing. It is therefore important to apply SHW Management of Organizational Change procedures on cost-saving programs as well as on investment and change programs.

A somewhat paradoxical challenge concerns the continued combined focus on cost reductions and better process control, e.g., by *lean* or *six sigma* programmes. These programmes help to reduce deviations in processes which is associated with a lower likelihood of accidents. In the short term, these programmes are good for SHW. But in the longer term, they may lead to less time for reflection, higher work pressure and neglect of opportunities associated with radical innovations. Instead of using lean methodologies to maximize efficiency, process optimalization methodologies and SHW can be integrated.

An often neglected example of drift to danger is associated with a pursuit of greater flexibility within teams. To increase *flexibility*, individual operators are often trained and formally qualified to perform work not only in their own discipline (e.g., mechanical) but also in other disciplines that are considered important for the team (e.g., electronics or ICT). Under normal circumstances, the operators will work in their own discipline, but now and then they may have to perform in the newly acquired disciplines. As they have little practical experience in the latter, they may not be really competent to do so, even though they are formally qualified. This increases the likelihood of human error and the associated SHW risks.

In the area of safety management, especially in the process industries, changes are known to require *management of change* (MoC) procedures. MoC for technological issues is in principle rather straightforward and can be dealt with rationally. However, the evaluation of management of organizational change (MoOC – e.g., significant changes in the organizational structure or in key personnel) is more complex and less straightforward. This implies the risk of incomplete or too optimistic risk assessment.

5.7.4 Promoting and Enabling Innovation and Change

Senior leadership should ensure that the opportunities implied by new developments, innovation and change are used not only to improve business processes but also SHW. This includes opportunities to promote and enable SHW in the early stages of development, design and procurement. It is vital to ensure that technical or organizational change or innovation always goes hand in hand with involving and developing people.

For leaders, managing organizational change is often a great challenge. How to motivate people and to overcome resistance to change? Research shows that it is possible to manage change processes successfully, even when it implies a reduction of personnel.[24]

Already when it is not yet decided how the aims of the change will be achieved, it is essential to inform supervisors and operational staff about the plan of change. First of all, it is necessary to clearly explain *why* the change is necessary, taking into account the psychological contract (see Table 5.2). Then it is important to be open about the principles that are to be followed, e.g., people will be involved in the development of the change (defining process and outcome). Knowledge and suggestions of the workforce will be seriously considered, though guarantees cannot be given. All relevant aspects of a change process will be taken into account, the focus will not be solely on more productivity. In this way, successful changes have been demonstrated in the areas of health[25] as well as safety.[26]

THE FIRST-MINUTE RISK ASSESSMENT[27]

Senior leaders are responsible for the design and functioning of the sociotechnical system, i.e., for the choice and design of technologies, the organization and their associated hazards and risks as well as for the people.

Proactive leaders carry out a first-minute risk assessment when a new business model or a new business opportunity is considered, i.e., before the socio-technical system is set up and choices regarding technologies and organization are made. This comprises the following questions:

- Is the new activity in line with our corporate and my personal values?
- Is there a (potential) conflict with fundamental labour and human rights?
- Is the new activity in line with or corporate strategy for Vision Zero?
- Is there a (potential) conflict with the expectations of key stakeholders (clients, authorities, etc.) regarding SHW?
- Is there a (potential) conflict with existing or expected SHW legislation?

The first time I saw a first-minute risk assessment in practice was in 1986. An AKZO research laboratory developed new peroxides for polyurethane production. The company regarded investments in the development of substances with mutagenic or carcinogenic properties as not a good business strategy. Each new substance underwent an AMES test.[28] Research and Development was only continued with substances without such properties.

5.7.5 Dealing Positively with Critical SHW Events

Dealing with the aftermath of serious SHW incidents is a relatively rare challenge for senior managers. A serious SHW incident causes a kind of crisis. It triggers reflection on the value of SHW for the organization and its people. It may lead to re-evaluation of the core values of the organization and its leaders. In this way, a fatal accident can be the start of an innovative programme to improve safety. Many senior leaders then state: 'this may never happen again in our organization'. In a similar way, a work-related suicide or cases of burnout, violence, bullying, work-related cancer or the sudden death of an employee after a stroke may trigger senior leaders to launch programmes for the prevention and promotion of better physical and/or mental health. Organizations not yet committed to Vision Zero regularly commit themselves to Vision Zero after such a tragic event. See also Section 9.6.

TWO CRITICAL MOMENTS TO SHOW PERSONAL COMMITMENT

It is easy to say that people are welcome to report SHW problems. It is also easy to say that it is OK to suspend work in case of serious risks. But reports on SHW problems always come at an inconvenient moment, while suspending work directly affects productivity. Showing personal commitment in both cases means that the reports or safety concerns are taken seriously, that those involved feel they are respected and that there are no negative responses to them. In both cases, it is also crucial that the leader is (to some extent) personally involved in the follow-up. When the critical situations are dealt with, positive feedback to the reporter or those who suspended the work is relevant as well.

A serious incident is an important event in the eyes of all employees. These are moments that really matter for successful SHW leadership. Does the leader really show empathy for those directly involved, including the family? Is there a genuine will to learn from the event to prevent reoccurrence? If the answer to one of these questions is no, this will feed scepticism among the employees: nice words that people matter are then contradicted by actual behaviour. Behaviour is much more convincing than words, and future messages espoused by senior leaders will be received with much more scepticism than before.

5.7.6 The Communication between Managers and SHW Professionals

Managers and SHW professionals often have different mental models concerning SHW and business processes. SHW professionals tend to emphasize cost reductions and SHW benefits realized through investing in SHW, while senior leaders are often primarily interested in the associated business benefits. As a result, communication about benefits and costs can be suboptimal. While senior leaders have to better understand the SHW benefits, SHW professionals have to learn to think and communicate about the business benefits. Senior leaders often have a strategic view of benefits and costs and are more open to embracing intangible benefits (like attractiveness in the labour market). Lower managers and supervisors focus more on benefits and cost reductions that can be expressed in monetary terms.

Against this background, it is useful for SHW experts to ask senior leaders their perception of the credible worst-case scenario relevant to their organization. What scenario has the potential to keep them awake at night? The responses can be surprising. A CEO of a chemical company did not hesitate and said: 'Three lost time incidents and we are out of business'. An occupational safety risk was for him primarily a significant business risk. Safety professionals would probably think of a much less likely scenario that would lead to a process safety incident with a major impact on the organization's environment. Another CEO responded that 'the prevention of any scenario that may lead to a serious accident is a top priority', and he explained that the impacts of a serious accident on business continuity, organization and brand reputation were his main concern. SHW can also be associated with serious liability risks, e.g., in the case of occupational illnesses like cancer.

SHW can seriously impact the organization's status in the labour market. High levels of turnover due to poor working conditions or lack of opportunities for personal growth may affect the organization's capabilities for production and innovation. This is especially relevant in case of a shortage of qualified workers in a booming economy. A less obvious business risk is when high levels of stress impede the creativity on operational level and undermine an organization's innovation strategy. Business and SHW risks are associated. In all these contexts SHW professionals should be aware that business risks are usually at least as important for senior leaders as SHW risks.

5.8 LIMITATIONS OF LEADERSHIP

Leadership is a main factor in improving SHW, but there are limitations to what a good SHW leader can accomplish. These limitations can be framed as *challenges*, but overcoming these limitations can be difficult and require endurance and self-confidence; even then successes may be partial only.

Leaders are often honoured and confronted with sky-high expectations. But sometimes it seems beyond their power to realize the changes needed. Employees may resist to the plans of their leaders or deny the problems for which the intended change is considered the solution. Leaders may be confronted with a struggle over conflicting (often incompatible) interests, ideas or preferences, with a lack of competence, outdated technologies or economic difficulties. Leaders have to deal not only with rational behaviour but also with personal interests, emotions, fantasies and fakes.

The first limitation is the competence and experience of the leader. Every leader has strengths, weaknesses and blind spots. It is very important that leaders are aware of their weaknesses and are part of a team that is able to compensate for their weaknesses and to raise discussion on issues that may be a blind spot for the leader in charge.

Senior leaders are not able to control the external organizational environment and can be confronted with all kinds of external pressures and changes, e.g., in economic developments, the labour market, in legislation, in expectations of specific stakeholders or the general public. But there are also internal factors that are difficult to control: within larger organizations, it is not uncommon to find subcultures. To a certain extent, any organization reflects the society or community it is based in. Values and habits in those communities are not always conducive to SHW. For example: discrimination is sometimes prevalent in communities or societies and is then likely also to play a role in the organization. The adoption of Vision Zero can be a good reason to start meaningful conversations on what the organization wants and how that relates to the societal norms in the community they are based in.

The ways of communication may differ among cultural groups, e.g., in many Asian cultures it is a taboo to say 'no', in other cultures it can be a taboo to refuse an order from somebody higher in the hierarchy or to raise questions about orders or instructions. In many European construction sites, personnel of a range of sub- and sub-sub-contractors are working, often originally, from other European countries with a range of mother tongues and only a limited understanding of the most common language used on site, which can be the language of the country where the organization is based in (e.g., the German language in Austria) or English. This hampers good communication and safe work.

There are also communities or sectors of industry where there is a history of adversarial industrial relations. This can easily lead to an *us versus them* culture, wherein leaders and workers mistrust each other, do hardly speak to each other, let alone listen to each other. This can lead to labour conflicts and easily escalate to strikes, etc. It can be very difficult to manage a break-through in an *us versus them culture*'.[29] In such cases, senior leaders will have to take the initiative – which they will do only with some hesitations. These initiatives will be received with suspicion. It is easy to fall back into the classic patterns, which are engrained in the mind of both leaders and operational staff and their representatives. This may even imply that workers are not really interested in safety at all, even when there are good reasons for improving safety and health. Trade unions may then see improvement in safety and health as a method to deprive them of the power to suspend work and shut down production in case of unsafe situations, which is important for profiling themselves to those they represent.

There are several reasons why exaggerated expectations of leadership regularly turn up. The first reason is that many leaders at the top love their power. They like to believe that it's in their power to control everything in their organization, even though they rationally know that there are many hierarchical layers, and each layer is a barrier to receiving relevant bottom-up information, as well as for disseminating information top-down. And yes, the CEO is accountable for what happens in the organization. That is probably why many operational people believe that the CEO is responsible for everything (good or bad) in the organization, thereby neglecting their own sphere of influence. Consultants are probably the group that is most outspoken about the importance of senior leadership. Management consultants often speak with self-admiration about their work with CEOs and speak with restrained admiration about their senior clients. An important reason is the self-interest of the management consultants. Working with the senior leaders helps to raise their status; for them, it is a selling point. The high status of the target group is reflected in the daily rates of the consultants.

Legislators also seem to believe in the omnipotent CEO. In several jurisdictions, they introduced manslaughter laws thereby making CEOs personally legally accountable and liable. The intention of these laws is, of course, to give senior managers a personal motive to actively promote safety and health and to do their utmost to prevent serious accidents and work-related ill health. However, the impact of such legislation seems ambiguous and questionable.

In the safety discipline a concept of *Work as Imagined* (WaI) versus *Work as Done* (WaD) has emerged, whereby WaI reflects the rules and procedures, and WaD stands for the reality in operational practice which is often decisive for safety. It makes sense to make a similar distinction between *Management as Imagined* (MaI) versus *Management as Done* (MaD). MaI reflects the management policies, the espoused values, the communications on the organization's website, etc. MaD stands for the visibility, the commitment and competence of managers and leaders and how they promote SHW in practice.

5.9 CONCLUSIONS

For Vision Zero organizations, it is vital to have committed and competent leaders. New leaders should be selected on their values and motivation for and their track record in SHW. Leadership is important for Vision Zero at all hierarchical levels. But committed senior leadership is vital to give direction to the development of the organization, for perseverance over longer periods of time, the development of a prevention culture and promoting and enabling proactivity and organizational learning.

Senior leaders are the personal face of the organizations. It is important for them to be in control. For information on SHW at the operational level, they depend on reporting systems, meetings with supervisors and operational staff on site. There are not many occasions for senior leaders to show personal commitment. Regular workplace visits are important in this respect and should be used to understand the challenges operational workers face and to show commitment.

The visible personal commitment of senior leaders is crucial for 'giving direction'. That is propagating the value of SHW, showing exemplary behaviour, involving people, giving time to discuss SHW matters and making resources available (time, money, knowledge, tools). Employers expect their employees to be committed to the organizational goals. Employees expect their employers to practice social justice. Their relationship, though unequal in terms of power and interests, can be a good one and work well. It can be useful to discuss the informal psychological contract to restore or develop mutual trust between management and workforce. Leaders should regard the people who work for them as a kind of extended family and take care of them in good but also in difficult times; that is good employership.

Motivated people should be able to make choices themselves and enjoy a reasonable degree of autonomy. This enables them to behave proactively and also facilitates their engagement. Leaders should give social support to individuals and teams.

Managers have power. It is a challenge to use their powers wisely and humbly. The powers of senior leaders should ensure that SHW is already addressed in the early stages of development, design and procurement and that problems encountered during production are fed back to those involved in the design stage. Leaders should take responsibility for the SHW of all members of the organization and actively promote SHW improvements. They have to ensure that necessary improvements are implemented timely. Be alert to upcoming SHW risks and actively promote organizational learning. Finally, it is important to create a climate wherein dialogues and discussions on a variety of ideas are welcomed. Leadership is the art of using power wisely.

Important leadership challenges discussed are (1) receiving adequate SHW information, (2) balancing SHW and productivity, (3) preventing drift to danger, (4) promoting and enabling innovation and change, (5) dealing positively with critical SHW events and (6) the communication between managers and SHW professionals.

There are limitations to what a leader can accomplish. Leaders are often confronted with unduly expectations. Leaders have their own weaknesses and blind spots and should have a team that compensates for them. Moreover, there are external as well as internal factors they cannot fully control.

NOTES

1 Text based on an interview with Peter Spit on 11 December 2024.

2 See: Zwetsloot G.I.J.M., de Bruin, T., (2022) Surviving relatives as stakeholders for corporate social responsibility and as leaders for meaningful safety improvement – A case study from the Netherlands, *Safety Science*, 157, 105927. Doi: 10.1016/j.ssci.2022.105927.

3 Bass, B.M., (1999) Two decades of research and development in transformational leadership. *European Journal of Work and Organizational Psychology*, 8 (1), 9–32. Doi: 10.1080/135943299398410.

4 See for an overview: Pilbeam, C., Doherty, N., Davidson, R., Denyer, D., (2016) Safety leadership practices for organizational safety compliance: Developing a research agenda from a review of the literature. *Safety Science,* 86, 110–121. More recent is the PhD by the very experienced safety consultant Victor Roggeveen. Roggeveen, V., (2022) *The influence of leadership on the prevention of safety incidents: on risk reduction, leadership, safety principles and practices.* Doctoral thesis. University of Leiden (NL). A very nice concise paper full of practical wisdom is that of Triss Kerin: Kerin, T., (2020) The importance of (process safety) leadership. *Journal of Loss Prevention in the Process Industries,* 66, 104220. https://doi.org/10.1016/j.jlp.2020.104220.

5 However, in some cases informal leaders can be cynical with respect to SHW and/or the management, then the organization is likely to suffer from an us versus them culture.

6 Locke, E.A., Latham, G.P., (2002) Building a practically useful theory of goal setting and task motivation – a 35 year Odyssey, *American Psychologist,* 57 (9), 707–717. Doi: 10.1037/0003-066X.57.9.705.

7 A Value case implies a broader perspective than traditional business cases. See: Van Scheppingen, A.R., Baken, N., Zwetsloot, G.I.J.M., Bos, E.H., Berkers, F., (2012) A value case methodology to enable a transition towards generative health management: a case study from the Netherlands, *Journal of Human Resource Costing and Accounting,* 16 (4), 302–319. Doi: 10.1108/14013381211317266.

8 Zwetsloot, G.I.J.M., (2017) Vision Zero: Promising perspectives and implementation failures – A commentary on the papers by Sherratt and Dainty, and Dekker. *Policy and Practice in Health and Safety,* 15 (2), 120–123. Doi: 10.1080/14773996.2017.1373998.

9 Compare with Goleman, D., Nevarez, M., (2018). Boost your emotional intelligence with these three questions, *Harvard Business Review,* 16 August, 1–7.

10 Bragdon, J.H., (2006) *Profit for Life. – How Capitalism Excels.* Cambridge, MA: SOL.

11 Bragdon, J.H., (2006) *Profit for Life. – How Capitalism Excels.* Cambridge, MA: SOL.

12 Rousseau, D.M., (2004) Psychological contracts in the workplace: understanding the ties that motivate. *Academy of Management Perspectives,* 18, 120–127. Doi: 10.5465/ame.2004.12689213.

13 Dabos, G.E., Rousseau, D.M., (2004) Mutuality and reciprocity in the psychological contracts of employees and employers. *Journal of Applied Psychology,* 89, 1, 52–72. Doi: 10.1037/0021-9010.89.1.52.

14 Beer, M., (2009) *High Commitment, High Performance – How to Build a Resilient Organization for Sustained Advantage.* San Francisco: Jossey-Bass.

15 Coyle-Shapiro, J.A.M., Kessler, I., (2000) Consequences of the psychological contract for the employment relationship: A large scale survey. *Journal of Management Studies,* 37(7), 903–930. Doi: 10.1111/1467-6486.00210.

16 Robinson, S.L., Rousseau, D.M., (1994) Violating the psychological contract: Not the exception but the norm. *Journal of Organizational Behaviour,* 15, 245–259. Doi: 10.1002/job.4030150306.

17 Gründemann, R., Huiskamp, R., (2008) Goed werkgeverschap (Good employership), in: Dhondt, S., Vaas, F., (2008). *Waardevol werk – Van arbeidskwaliteit naar sociale innovatie* (*Valuable Work – from Quality of Work to Social Innovation*). Den Haag: Lemma. pp. 171–188.

18 Griffin, M.A., Parker, K., Mason, C.M., (2010) Leader vision and the development of adaptive and proactive performance: A longitudinal study. *Journal of Applied Psychology,* 95 (1), 174–182. Doi: 10.1037/a0017263.

19 Griffin, M.A., Parker, K., Mason, C.M., (2010) Leader vision and the development of adaptive and proactive performance: A longitudinal study. *Journal of Applied Psychology,* 95 (1), 174–182. Doi: 10.1037/a0017263.

20 Zwetsloot, G.I.J.M., (1994) *Joint Management of Working Conditions, Environment and Quality; in Search of Synergy and Organizational Learning.* Amsterdam: NIA.

21 Hawkins, P., (2005) *The Wise Fool's Guide to Leadership.* Hants, UK: O-Books.

22 Mustard, C., Yanar, B., (2023) Estimating the financial benefits of employers' occupational health and safety expenditures. *Safety Science,* 159, 106008. Doi: 10.1016/j.ssci.2022.106008.

23 Rasmussen, J., (1997) Risk management in a dynamic society: a modelling problem. *Safety Science,* 27 (2–3), 183–213. Doi: 10.1016/S0925-7535(97)00052-0.

24 Zwetsloot, G.I.J.M., Gort, J., Steijger, N., Moonen, C., (2007) Management of change: Lessons learned from optimising manning staff reductions in the chemical process industry, *Safety Science,* 45, 769–789. Doi: 10.1016/j.ssci.2006.08.028.

25 Van Scheppingen, A.R., de Vroome, E.M.M., Ten Have, C.J.M., Zwetsloot, G.I.J.M., Van Mechelen, W., (2014) Inducing a health-promoting change process within an organization: The effectiveness of a large-scale intervention on social capital, openness, and autonomous motivation toward health, *Journal of Occupational and Environmental Medicine*, 56 (11), 1128–1136. Doi:10.1097/JOM/0000000000000299.

26 Zwetsloot, G.I.J.M., Gort, J., Steijger, N., Moonen, C., (2007) Management of change: Lessons learned from optimising manning staff reductions in the chemical process industry, *Safety Science*, 45, 769–789. Doi: 10.1016/j.ssci.2006.08.028.

27 Many organizations expect their workers to carry out a last minute risk assessment (LMRA) every time before they start their activities and provide guidelines how to carry out that assessment. Contrastingly, first minute risk assessments are usually less systematically planned and carried out and are often not considered to be part of the SHW management system.

28 The Ames test is simple, quick and cheap method that uses bacteria to test whether a chemical has mutagenic and probably also carcinogenic properties. Ames, B.N., Durston, W.E., Yamasaki, E., Lee, F.D., (1973) Carcinogens are mutagens: a simple test system combining liver homogenates for activation and bacteria for detection. *Proceedings of the National Academy of Sciences of the United States of America*. 70 (8): 2281–2285. Doi: 10.1073/pnas.70.8.2281.

29 Schaap, L., (2021) *Brand in Amsterdam – Hoe een verziekte overheidscultuur kan blijven bestaan* (*Fire in Amsterdam – How a Screwed Up Governmental Culture Can Persist*), Amsterdam: Business contact.

6 From Traditional to Proactive Prevention and Promotion

Vision Zero is based on the belief that all (serious) work-related accidents and diseases are preventable. Prevention is therefore a key concept for Vision Zero and more in general for addressing *the human toll of work* (and the associated material damages and financial losses) as described in Chapter 1.

Accidents and work-related diseases have usually a long history. In case of diseases that is often long-term chronic exposure to certain health hazards. In case of accidents that can be the implicit acceptation of reduced safety margins or the malfunctioning of certain safety arrangements. Such underlying causes lead – always in hindsight – to sayings like *the accident was waiting to happen.* Usually, there are plenty of possibilities for timely prevention. But the concept of prevention is used in different ways which can be confusing and may give rise to question marks about its usefulness. Therefore, we first analyse the present confusion about the concept of prevention and the limitations of traditional prevention. The focus is thereby on the hierarchy of prevention measures and the distinction between primary, secondary and tertiary prevention. For Vision Zero, prevention has to be redefined in a proactive and strategic way: *proactive prevention.*

6.1 THE TRADITIONAL HIERARCHY OF PREVENTION MEASURES

There is not a broadly accepted definition of prevention. In the ISO standard 22316 for security and resilience,[1] the following definition is given for prevention (of hazards and threats): *measures that enable an organization to avoid, preclude or limit the impact of an undesirable event or potential disruption.* In the area of occupational safety and health often references are made to the hierarchy of prevention measures. See Figure 6.1.

The *hierarchy of prevention measures* is well-known, though there are several variations. Here we refer to a version that is widely used. It comprises five levels or steps. The idea is simple. Measures high in the hierarchy have a more effective preventive impact than those low in the hierarchy and should be considered seriously before applying measures that are lower in the hierarchy.

The highest priority is the *elimination of hazards*: the total removal of a specific hazard makes it impossible for the hazard to cause ill health or accidents.[2] When a

DOI: 10.1201/9781003621225-6

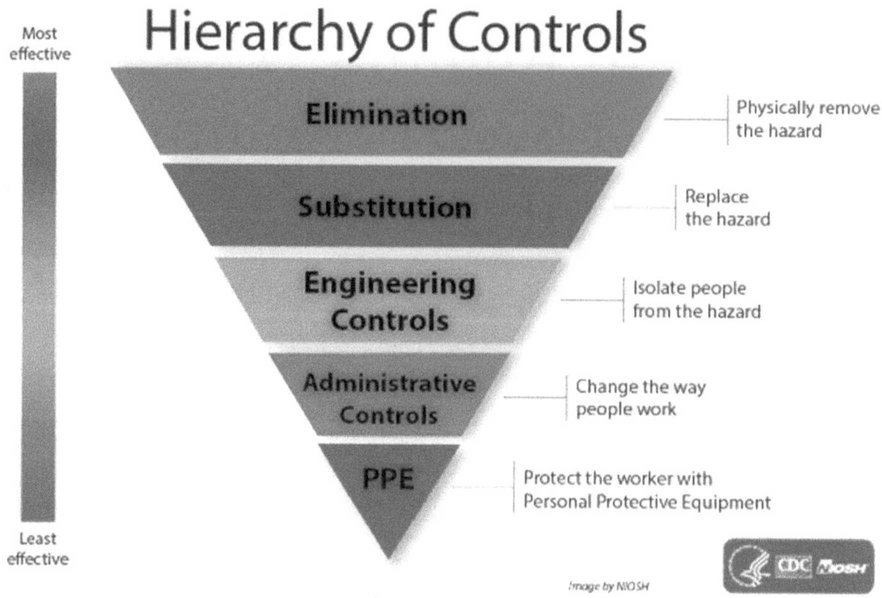

FIGURE 6.1 The hierarchy of prevention measures. (This model is used by NIOSH, see www.cdc.gov/niosh/hierarchy-of-controls/about/index.html (Accessed 28 November 2024) and also presented by EU OSHA in the OSH Wiki: https://en.wikipedia.org/wiki/Hierarchy_of_hazard_controls (Accessed 12 October 2024).)

hazard is removed, its threat to SHW is eliminated. It is to be noted that the eliminated hazard no longer requires additional control measures (e.g., training, monitoring, record keeping or Personal Protective Equipment, PPE) and their associated costs. These benefits of elimination remain relevant throughout the life time of the production process. An example is the elimination of manual lifting with its associated physical burden by automatic lifting.[3]

The second priority is *substitution:* it means replacing a hazard with one that is less dangerous, e.g., when a chemical powder is replaced by pellets (less exposure to dust) or when a highly flammable material is replaced by a less-flammable material.

The third priority is *collective engineering controls*: The remaining risks are reduced collectively by physical means, e.g., equipment that makes a lot of noise is located in a special room that is nobody's workplace.

The fourth priority regards *administrative controls:* organizational measures that aim to reduce exposure to the hazard by adherence to rules, procedures, protocols and/or instructions. These collective administrative controls depend on human behaviour. This makes the risk reduction less certain compared to the higher priorities. Administrative controls require education and training as well as adequate supervision implying ongoing costs.

The fifth and lowest priority in the hierarchy is *Personal Protective Equipment (PPE):* PPE protects the individual bearer. According to the hierarchy they should be used only when the other options are impossible or not feasible. This is often the case for maintenance activities. To be effective PPEs must be well-chosen, well-fitted to the individual, maintained correctly and worn consequently. Some PPEs are a nuisance to wear, e.g., respiratory protection, while this is not the case for hard hats or safety shoes. The nuisance to wear can often be reduced by fitting them to the individual; this makes it more likely that they will be worn when necessary. Like the administrative controls, the use of PPEs requires education and training and supervision and implies ongoing costs.

The measures at the two lowest levels require that they are understood, adopted and carefully and consequently applied by the individual workers. Not a very solid basis for prevention.

6.1.1 The Limitations of the Hierarchy of Prevention

The hierarchy of prevention was developed by safety engineers and occupational hygienists and is most useful for dealing with technological hazards. However, the hierarchy is not applicable to dealing with psychosocial hazards (e.g., conflicts at work or job uncertainty): elimination, substitution or collective engineering controls are not an option for psychosocial hazards. Moreover, the hierarchy refers to individual hazards, while in practically all workplaces there are several hazards simultaneously.

It is quite remarkable that important factors such (re)design of the work organization, leadership, assurance, education and training and standardization are not included in the hierarchy of measures, though these can be quite relevant for SHW prevention. The use of information technology is not mentioned either. Yet, innovative information technology is implying a lot of opportunities for prevention (as well as introducing new hazards and risks) for SHW.

There are several difficulties in how the hierarchy of prevention is applied in industrial practice. The options of elimination and substitution are often too easily disregarded. These are production-integrated measures that require adaptation of operational processes, technologies, or work organization. This makes their implementation complex; consequently, these options are often regarded as too costly or otherwise not feasible. That is why in practice the focus is strongly on the two lowest priorities: administrative controls and PPEs. These measures do not require the need to reconsider – and perhaps change – the production processes, the work organization or the workplace. These are added-on measures that can be applied in almost any situation without much consequences for the production process or the work organization. They do not require serious investments, so they are relatively cheap to introduce. This is probably why the measures at the two lowest levels are often preferred. This implies that in practice often less-effective measures are preferred. As Trevor Kletz stated about prevention measures: '*Following traditional methods we become more and more efficient at things we should not be doing, except as a last resort*'.[4]

All too often, SHW experts are not expected to interfere with the production facilities or the working processes, which would require innovation or organizational change. That limits their possibilities for contributing to more effective prevention. A more strategic fulfilment of the SHW professions is desirable.

Administrative measures are in itself not a bad idea, but they are often associated with lack of flexibility and autonomy, and an overkill of bureaucracy (paper safety with an abundance of rules, procedures, protocols and instructions). Procedures are not necessarily instructions that prescribe exactly what and how things have to be done. They can also have a form wherein people are encouraged to think about certain aspects before or when they take action.[5] When you play football or chess, you need rules. Rules create clarity about what is allowed or not. Nevertheless, both in football and chess, there are almost endless possibilities for playing a match or game. The rules create a framework, wherein creativity and sensemaking can flourish. Without such *rules of the game* uncertainty about the context may hinder creativity. However, when rules are imposed top-down, that frustrates creativity. To avoid procedures that hinder operational processes, it is important to develop them in a participative manner and involve the people to whom the envisaged procedures are addressed.

Finally, the hierarchy suggests that the involvement of operational staff in prevention is limited to acting in compliance with rules and procedures and using PPEs as prescribed. That leads to programmes of behaviour-based safety that aim to control the behaviour of the workers and to reduce human errors. That does not lead to programmes that stimulate the active involvement of workers in developing measures in line with the higher levels of the hierarchy.

We can conclude that the traditional hierarchy of prevention measures has significant conceptual and practical limitations. It is not helpful for managing wellbeing, important types of prevention measures are not included and the hierarchy is often implicitly used to justify the choice for the easier and less effective prevention measures. The hierarchy of prevention clearly has too many limitations to be useful for implementing Vision Zero.

6.2 PRIMARY, SECONDARY AND TERTIARY PREVENTION

Prevention in the area of occupational health and wellbeing has many things in common with the prevention in the area of occupational safety but is different in one important aspect: 'deviations from normal', which often precede accidents, are always unplanned and unintended. But the exposure of workers to conditions that impact worker health is usually chronic and in itself well-known. It is usually implicitly accepted as a side effect of planning the operational process (e.g., exposure to chemicals, noise, vibration, the need to lift heavy loads physically, high-speed repetitive tasks, etc.). But the potential long-term consequence of such exposures – which may cause irreversible health damage – is often a neglected consequence.

While the safety and industrial hygiene professionals often refer to the hierarchy of prevention measures, health and wellbeing professionals usually refer to another type of hierarchy: the distinction between primary, secondary, and tertiary prevention.[6] These prevention types refer to measures taken in different stages and degrees of (physical and mental) health problems.[7]

Primary prevention refers to measures that prevent ill health. This concerns a broad variety including all measures mentioned in the hierarchy of prevention measures, from the elimination or reduction of hazards to the use of PPE. Vaccination against infectious diseases is also a relevant form of primary health prevention. A psychosocial example is the reduction of stress risk by organizational (re)design, creating a balance between job demands, control and support.[8]

Secondary prevention refers to measures to avoid progress in ill health and to strengthen individual resilience. Secondary prevention is relevant in situations where (chronic) exposures exist or chronic health problems have already arisen and the aim is to better cope with ill-health or its causes. Through education or training, people learn to recognize critical situations and relevant symptoms, to increase self-awareness and to improve their coping skills. Job crafting is another example.

Tertiary prevention refers to activities aiming at the recovery and return to work of people suffering from ill health. It comprises counselling activities or educational activities on how to cope with work-related strain.

6.2.1 Limitations of the Differentiation between Primary, Secondary and Tertiary Prevention

The distinction and implicit hierarchy of primary, secondary and tertiary prevention is useful for the medical professions as well as for psychologists. They deal mainly with patients who already have health complaints. Primary prevention is then not a very relevant option. Different from most people, medical professionals may regard measures taken when people already suffer from ill-health as (secondary or tertiary) prevention. When we compare the hierarchy of prevention measures with the primary-secondary-tertiary distinction, it is remarkable that even the lowest priority in the hierarchy, wearing PPE, is a form of primary prevention. This clearly illustrates the confusion about the concept of prevention. Vision Zero is mainly focused on primary prevention, which does not imply that secondary and tertiary prevention (or emergency preparedness) should be neglected.

Effective prevention measures protect people from accidents or ill health. Workers thereby have their own responsibility and should be actively involved in creating safe and healthy working conditions. For organizations, it is important that workers know how to deal with the SHW risks relevant for their job, while being protected against irreversible damage to SHW which they cannot influence themselves.

6.3 SHW Promotion

Promotion of SHW has a long history. Campaigns for the promotion of safety and physical health have existed for more than a century; initially, they were predominantly aimed at increased awareness of risks, and the promotion of desirable safe and healthy behaviour. The concept of workplace health promotion is more recent and was initially only focused on the promotion of healthy life styles, e.g., physical activity, nutrition and discouraging unhealthy lifestyles like smoking or the use of alcohol and drugs. But for a few decades, the focus is also on improving the work organization and the work environment, on promoting active participation, encouraging personal development and addressing mental health issues as is clarified by the Luxembourg Declaration on Workplace Health Promotion in the European Union, which was launched initially in 1997.[9] Workplace health promotion is usually regarded as a form of secondary prevention, as its focus is on dealing with (existing and potential) health risks, i.e., health risks that remain after primary prevention measures are taken.

There are four main reasons for the growing attention for and importance of SHW promotion.

1. Due to an ageing working population, welfare diseases (obesity, diabetes, etc.) have entered the workplace. This implies that the employer has an increasing interest in promoting healthy behaviour in the workforce, also beyond the workplace, e.g., to promote life style issues like physical exercise, good nutrition and sufficient relaxation, and to discourage smoking and the use of alcohol and drugs. The attention for life styles may, however, distract from healthy work styles and a healthy organization of work.
2. The concept of safety promotion has recently gained popularity for developing individual and organizational capabilities to deal with upcoming or unexpected risks that emerge due to complexity; this is especially the case in the so-called 'new view' of safety.[10] Production processes have become much more complex than ever before, due to outsourcing, globalization, rapid technological developments and high degrees of specialization. Systems with greater complexities imply more uncertainties; this limits control measures and makes risks less predictable. The likelihood of surprises increases. In complex systems, not all relevant SHW scenarios are known, and controlling these scenarios may give rise to high costs or an unpractical number of rules and prescriptions.
3. Many jobs have become temporary, making job changes part of almost everybody's career. This implies a decreasing interest of the employer to invest in the health or development of their employees, and a decreasing loyalty of the employees towards the organization they work for. This implies a decreasing interest of both employers and workers in prevention measures. As a result, it becomes more important that working people are well-prepared to understand and deal with the SHW risks they may encounter in the workplace.[11] This increasingly underlines that SHW promotion is an indispensable complement to prevention.
4. The relevance of SHW promotion is augmented by the increasing diversity of people at work in terms of generations, gender, race, religion, mother language, cultural backgrounds, values, etc. This implies that a 'one size fits all' approach is not meaningful and that a variety of behavioural responses to riskfull situations are possible.

Promotion is an important complement to SHW prevention for dealing with the remaining and upcoming risks and developing excellence in SHW. However, it cannot be a substitute for effective prevention measures at the level of the (socio-technical) system, it is complementary to such system approaches. It is essential to understand that there are several aspects of human behaviour which are relevant for prevention but cannot be 'controlled' and only can be enabled or encouraged. Examples are: commitment, creativity, intrinsic motivation, learning, ownership and proactivity.

In this book, the concept of promotion applies to those activities that enable or encourage prevention in a proactive way, which include increasing human and organizational capabilities as well as promoting responsible behaviour and creating a prevention culture.

6.3.1 Limitations due to System Level

Decision-making about prevention is influenced by the context, e.g., time, space, economic situation or management style. Prevention can be focused on various system levels, involving differences in space and time: e.g., the equipment, the work place, the individual, the team, the department, the factory, the firm as a whole, the value chain, the sector or the country. The system level chosen influences the need and options for prevention. It is important not to define the system too narrowly. Wider system boundaries imply more options for effective prevention. When considering prevention, it is always important to reflect critically on the (often implicitly chosen) system levels.

Timing is often important. Long-term ambitions (like Vision Zero) may conflict with short-term interests. There are many examples where short-term solutions are counterproductive for long-term solutions/ambitions (e.g., problem solving instead of innovation). Considering long-term implications is very important for chronic or persistent SHW issues and Vision Zero.

There is consistent evidence that occupational safety interventions should foster safer working environments, machines, tools, work organization and working conditions rather than solely focusing on how workers can mitigate the risks.[12] In companies, it is important that lessons learned in a specific department might also be useful in other departments.

6.4 SUMMARIZING THE CONFUSION ABOUT TRADITIONAL PREVENTION

The hierarchy of prevention does not match logically with the concepts of primary, secondary and tertiary prevention. Both the hierarchy of prevention measures and the differentiation between primary, secondary, and tertiary prevention have serious limitations.

Both the hierarchy of prevention and primary and secondary prevention focus on operational measures that eliminate, reduce or control hazards and risks in concrete situations. It can therefore hardly be a surprise that prevention is not a strategic priority for most organizations. This is usually only the case for safety in the so-called 'high risk' sectors (nuclear, (petro-)chemical, air, rail, etc.) where the impact of a disaster can lead to fatalities and have a high impact on society, may lead to loss of reputation and the 'license to operate', apart from significant financial consequences.

SHW problems occur, and they have to be solved. But it is confusing to call problem-solving prevention. As long as the focus is on identifying and solving existing SHW risks, the resulting policies will dominantly be reactive, focused on short-term results and compliance. Hollnagel even stated that prevention is basically reactive and has limited potential,[13] but he focuses strongly on the stage of production and maintenance, neglecting prevention in the earlier stages of conceptualization, design and planning.

It is easy to see that the traditional hierarchy *of prevention measures* as a concept for prevention is ambiguous and confusing and tends to be reactive and operational; it is usually neither proactive nor ambitious. Also, in the primary, secondary and tertiary prevention concepts, secondary and tertiary prevention are in fact accepting implicitly undesirable situations, such as exposure to noise, chemicals or stress. So, both concepts do not sufficiently support a long-term strategy for prevention. That is why there is a need to redefine the concept of prevention in a proactive and strategic way, relevant to all three aspects of safety, health and wellbeing at work.

6.5 PROACTIVE PREVENTION

Given the confusion around the concept of prevention, there is clearly a need to redefine prevention in a proactive way. When we want to redefine prevention more proactively, we can find some inspiration in the definitions of prevention in the areas of total quality management and sustainability.

In the area of total quality management prevention is defined as everything that is needed *to do things right from the start* (see also Section 4.4); this implies that the timing of prevention measures is crucial. Defined that way, prevention includes e.g., leadership, planning, design reviews, training and education, procurement, etc. Prevention is thereby distinguished from 'monitoring' (every activity focused on assessing or measuring relevant situations), 'correction' (problem-solving) and those 'dealing with failure' (mitigating or compensating for the impact of non-quality).[14] Of course, excellent quality management does not only require prevention but also monitoring, correction and dealing with failure. But investments in prevention imply that the costs for monitoring, correction and dealing with failure can be reduced; in this way prevention in quality pays. Or as Crosby framed it: Quality is free.[15]

In the area of sustainability, the term prevention is used as part of the combination of *waste prevention*[16] or *pollution prevention.*[17] Both emphasize the importance of prevention as an integrated process to reduce *the generation* of waste or pollution. Measures to mitigate *the effect* of waste or pollution, e.g., by filtering out toxic components are not regarded as prevention, but instead as (end of pipe or added-on) *control measures*. Investments in environmental prevention imply a decreasing need for control measures and a reduction of the costs thereof. That is communicated with the well-known slogan Pollution Prevention Pays.

In Section 4.5, proactivity was already discussed in more detail. The core of proactivity comprises: (1) Making things happen, (2) anticipating and preventing problems as early as possible and (3) seizing opportunities for improvement.[18] The consequences for proactive prevention are: (1) A shift in focus from SHW prevention only in the production phase to also include the early phases of development, design and planning; (2) Identifying remaining and upcoming hazards and risks in the production phase as early as possible followed by measures to reduce them. (3) Using innovation and change as drivers for prevention (rather than seeing them mainly as a source of new risks).

To align those strict definitions of prevention used in the areas of total quality management and sustainability with the three dimensions of proactivity (see Section 4.5), the definition of proactive prevention should include timing. An important characteristic of proactive prevention measures is that they are taken as early as possible (in the design stage as well as during production). Proactive prevention should also focus on measures integrated into the production process itself thereby reducing the need for added-on control measures, monitoring, correction, etc. This implies that the choice of technologies and organizational arrangements is at stake and that senior leaders should take the lead in proactive prevention. Finally, innovation and change should be used to generate opportunities for proactive prevention (rather than seeing them mainly as a source of new risks).

Proactive prevention then includes all activities *that help to prevent the generation of SHW hazards and risks* (but not *the control* of these hazards and risks). These are organizational or technological measures integrated into the production processes or work organization.[19] Proactive prevention is fully in line with the quality principle of 'doing the right things right from the start' (see Section 4.4) and often implies worker-friendly innovation.

Preventive Control is when production and work arrangements are given, and added-on measures (e.g., engineering controls, procedures, PPE), or minor adaptations of work and production processes (e.g., job crafting) are taken to reduce and control the risks. Preventive control measures can be regarded as added-on corrections to the prevailing production and work arrangements. Both proactive prevention and control measures can prevent accidents, incidents and ill-health; the added-on nature of control measures, however, implies ongoing activities and costs to make sure the preventive control measures are still working as intended. It also implies the risk that they might fail (e.g., that filters for the purification of air, or procedures to control the behaviour of workers are not, or not any longer, as effective as expected).

A strong *prevention culture* enables and encourages proactive behaviour of individuals and teams aiming at proactive prevention complemented by preventive control.[20] The stronger the prevention culture, the stronger the focus on proactive prevention, see Chapter 7.

Proactive prevention will never lead to the elimination of all SHW hazards and risks, some rest risks will remain. Moreover, the future is unpredictable and in today's changing world it is not possible to ensure that no (new) SHW issues will ever arise. It therefore remains necessary to be alert to upcoming and emerging SHW hazards and risks (through alertness, risk assessments, monitoring or surveillance). Then, action can then be taken, preferably as early as possible. Emerging or newly assessed risks can be dealt with in a proactive preventive manner or through preventive control. Short-term economic feasibility will then often be in favour of preventive control measures, while a long-term perspective will often lead to the preference for proactive prevention. The better organizations are in proactive prevention, the less costs for control measures, monitoring and mitigating activities have to be made. Proactive Prevention Pays!

6.5.1 Proactive Prevention in the Early Stages of Development and Design

Research on the causes of process safety risks showed that poor design and poor validation of equipment are the most important causal factors of incidents in the oil and gas industry.[21] Kletz advocated the positive tone of voice:

> It is better to say an accident can be prevented by better design than to say it was caused by poor design. While people are willing to improve, they do not like to admit a poor performance as that is often associated with shame, blame, accountabilities and liabilities.[22]

In the high-risk sectors, the principle that prevention should start in the early phases of development, design and planning is broadly accepted by industry, scientists and legislators. Yet, it is remarkable that this principle is not yet generally accepted in many other sectors of industry for SHW prevention. Then, prevention is focused on the stages of production and maintenance and little serious attention is given to the earlier stages of development and design. But when prevention starts in the phase of production and execution of work processes, it tends to be reactive: many important decisions are then already taken. When insufficient attention is given to SHW in these early stages, SHW issues are very likely to arise and the organization and its people have to deal with problems already created. Positively formulated: the best options for creating good SHW are found in the early stages of the life cycle of production and work arrangements.

Research at the Saïd Business School showed that often projects don't go wrong; they already start wrong. Typically, top management decides early on a budget and schedule for the project before either can be verified.[23] This then causes financial constraints and time pressure. That triggers all kind of problems during development or construction, including SHW issues, e.g., high work pressure, hasty work, overwork, selecting contractors and technologies on budget rather than quality, etc. It causes teething problems in the production stage, and unnecessary SHW challenges associated with high costs in the stage of production and maintenance. It implies higher total costs of ownership during the life cycle of the production facility. A good start is a slow start with time to think things through.

To make a good start it is necessary to train or educate people (leaders and operational staff) before they start their job, or before new activities evolve. The education and coaching of young employees deserve special attention too as it may form their attitudes and habits during their career.

An important strategy to be early is to integrate SHW into decision-making processes regarding business processes, the organization of production and work, and the early (conceptual) stages of the design and planning. In these early stages, a variety of technological and organizational options should be considered, and after an assessment of the associated hazards and risks the most worker-friendly options can be selected. The use of a first-minute risk assessment (see Section 5.7.4) is then appropriate. It is the duty of senior leaders to make sure that SHW is addressed in the early stages and that the outcomes are integrated in the decision-making. This

TABLE 6.1
Activities relevant for proactive prevention in the early conceptual and design stages

SHW in research and development
SHW in design and design reviews
SHW in the selection of business partners
SHW in procurement and supplier evaluations
SHW in marketing and acquisition
SHW in the (re)design of organizational arrangements, reorganizations and work processes
SHW in planning of work and production
SHW in selection of new personnel (at all hierarchical levels) and contractors
SHW in training and education (of managers and workers) before the work begins
SHW as integrated part of professional development and professionalism
SHW strategy and management systems that focus on the early stages of design and planning, not only the production phase
Monitoring technological, organizational and societal developments outside the organization and assessing opportunities for SHW improvement
Feedback from SHW problems and lessons learned in the later stages to those departments or people involved in the early stages of decision-making and design

requires strong leadership. The positive impact on SHW and productivity will be achieved as long as production is running. This is compatible with the long-term perspective and ambitions of Vision Zero.

In these early stages, it may be necessary to go beyond organizational and professional boundaries. Construction firms will need to cooperate with architects and project developers. Manufacturing companies need to cooperate with suppliers, etc. From the perspective of architects or suppliers, it implies that their organizations have a responsibility to look forward to anticipate future consequences and do their best to reduce SHW risks in later stages of the life-cycle. Table 6.1 shows examples of activities relevant to proactive prevention in the early conceptual and design stages.

6.5.2 Design Thinking

Design thinking is complementary to management thinking, and it departs from an open mental model wherein opportunities are systematically and creatively explored. Design thinking is characterized by (1) focusing on an agreed aim, (2) gathering all possible options to achieve the goal, even when some options seem ridiculous at first sight (3) choosing from the options the alternative that is best with a view on the life cycle of the production and/or product or service. A good understanding of design thinking is vital for innovation and for the integration of SHW in the early conceptual and design stages. It is important to involve all relevant stakeholders in the process of generating alternative options and selecting the best option. In a work setting this implies involving workers or their representatives in developing or choosing the best available technology, work organization, design of workplaces or procurement of software and tools.

A vital resource for successful design thinking as well as for taking long-term impacts into consideration is *time for reflection*. Time is not only necessary to prevent projects from going wrong from the start, but it is also important in the stages of design and production. As mental models play a crucial role in thinking and decision-making, it is necessary to now and then take a *mental distance* from the ongoing issues of every day work. Only then reflection on the existing mental models and their long-term implications, and possible alternatives can be successful. Time is also needed for a fruitful dialogue with stakeholders involved, including those from other disciplines or departments; dialogues are very helpful to consider the options from several points of view and to encourage mutual learning. Good dialogues require a culture of trust, active participation and careful listening to each other.

6.5.3 Precaution

While prevention of accidents is focussing on sudden events, for health and well-being the focus is on chronic exposures that can have a long-term impact. In the area of health and wellbeing warning signals often relate to limited scientific evidence about the long-term impact of chronic exposure to certain hazards, like toxic substances.[24] Unfortunately, the value of risk assessments is limited when there is a lack of scientific basis for long-term and uncertain effects. In such cases, proactive prevention requires a variation of 'better safe than sorry'. *The precautionary principle* applies when scientific evidence has found a plausible serious risk but conclusive evidence is not (yet) available. It implies that in case of threats of future serious or irreversible damage, lack of full scientific certainty (i.e., in case of uncertainty, complexity or ambiguity) should not be used as a reason for postponing prevention measures.[25]

Precaution in the area of SHW is clearly relevant for prevention in cases of the usage of carcinogenic, reprotoxic or mutagenic chemicals. For these long-term irreversible health effects, full evidence is only available for a minority of chemicals. Then prevention can often not wait till more scientific evidence is available. This is why the European Union adopted the REACH regulation,[26] i.e., the legislation on the registration, evaluation, authorization and restriction of chemicals. The precautionary principle might also be relevant for other new types of health risks, e.g., for working with nanomaterials, or exposure to microplastics.

Decision-making under conditions of uncertainty, complexity and ambiguity can also be relevant for safety. Uncertainties in safety are especially important in the case of radical innovations, whereby totally new technologies are developed and applied, or whereby technologies are applied under novel or very extreme conditions[27] or where operational uncertainties may create events leading to very serious, even disastrous consequences (e.g., multiple fatalities[28]).

There is growing evidence that psychosocial stressors can lead to 'hard' effects like increased mortality due to coronary heart disease. Another example is that structural long working hours (>55 hours per week) are now considered the main cause of work-related death globally.[29] These examples show that the precautionary principle is increasingly relevant in the area of mental health and wellbeing as well.

6.5.4 Proactive Prevention and Work Organization

The work organization is often neglected but deserves dedicated attention as it has consequences for wellbeing but also for health and safety. The importance of choices regarding the organization of work was already shown by Taylor in the early twentieth century, who advocated specialization and simplification of tasks. Later the disadvantages of specialization and monotonous jobs for health and wellbeing have come to the foreground. Work organization design became the central issue in the socio-technical school initiated in 1951 by Trist and Bamforth of the Tavistock institute.[30] The researchers of this school demonstrated that by giving workers control over their work, improvements in team work, productivity and health, safety and wellbeing could be combined. Nowadays it is clear that taking psychosocial hazards into account in the design of the work organization has a great potential to improve not only wellbeing but also health and safety, productivity and innovation.

6.5.5 The 12 Principles of Worker-Friendlier Design

The technologies and work organization chosen have many consequences for SHW. Trevor Kletz was the first to promote inherently safer production processes:[31] Production processes made safer by their very nature and not because equipment has been added to make it safer, or complementary organizational arrangements.[32] The (re)design process should not remain the responsibility of the engineers only: senior leadership is crucial.[33]

From a Vision Zero perspective, proactive prevention should include inherently (physically and mentally) healthier design.[34] We call this broader approach *worker-friendly design*[35] *(WFD)*. It implies that attention is given to technological and organizational choices made in the development, design and planning of business models, products and services, work organization, work processes, workplaces and tools. The focus is thereby on the elimination or significant reduction of SHW hazards and risks. Senior leaders can and should lead and support the challenge of *worker-friendlier design,* which is associated with technological as well as organizational innovation. WFD implies the application of a limited series of 12 design principles which should be systematically integrated into the development and design.

Worker-friendlier (re)design can help to solve persistent SHW problems (which have proven to be difficult to solve with more traditional SHW approaches). Persistent or repetitive problems may require the redesign of the work organization and/or the introduction of alternative technologies in the workplace.[36]

Leaders who promote WFD aim at removing barriers to make the necessary choices. The barriers can be manifold: mental models, habits, standards, legislation and existing contracts and licenses. It is important to acknowledge and communicate the (strategic, economic and SHW) benefits of worker-friendlier production. It is important to minimize defensive reactions such as *not invented here, not in my turn*

TABLE 6.2
The 12 principles for worker-friendlier design

Principles for Worker-Friendlier Design	Explanation/Example
Purpose	Work is important to people, especially when their work is experienced as meaningful. What is the purpose of production and work? Can we create more societal or social value? Maximize meaningful tasks and opportunities for personal growth. It may help to ask the questions: Why do we do this? What could we do instead? How do we generate maximum societal value?
Be early	To be early with SHW implies to integrate it in the earliest stages of decision-making and design (the first-minute risk assessment). In the early stages, SHW measures can easily be integrated, at relatively low costs while generating SHW and business benefits for the rest of the life cycle of the production unit. But the principle remains important in later stages. During production, it implies recognizing early warnings, often in the form of weak signals, and taking measures before accidents or illnesses can occur, see Section 9.4). Proactivity implies early anticipation of potential problems.
Precaution	The application of the precautionary principle is actually a variation of minimization. In case of scientific uncertainty, ambiguity or complexity, about the (long-term) impact of a hazard, but indications of potentially irreversible future damage (e.g., when introducing carcinogenic, mutagenic or reprotoxic chemicals) precaution implies elimination or minimization of the use of the hazardous component, or to use them only under strict conditions and extensive monitoring. When more evidence about the risks becomes available, this may lead to less stricter conditions.
Design for people using 'human factors'	Adaptation of technology and work to the worker is a basic ergonomic principle. It implies taking into account *human factors*, i.e., adapting technology, organization, work processes, tools, etc. to the needs, possibilities and limitations of the worker. It strongly reduces physical or mental load as well as the likelihood of human error. The ergonomic principle often coincides or strengthens the principle of 'simplification' from the perspective of the worker. The *design for all* principle implies that it is possible for the workers to adapt the workplace or tools to their needs (e.g., to adapt the height of a desk) and that the work can also be adapted to people with a (physical or mental) disability.

TABLE 6.2 (Continued)
The 12 principles for worker-friendlier design

Principles for Worker-Friendlier Design	Explanation/Example
Participation	It is important to involve the future users (workers) in the design of the work organization, work processes, tools and workplaces. Involvement in the (re)design or the development of a solution creates ownership. Participation means making use of the tacit knowledge of the workforce and often is the simplest way to ensure that the ergonomic principle of adaptation to the worker is applied. It is often the best guarantee that the solution will work in practice. In the case of new technologies, it is often useful to involve other key stakeholders as well, to ensure that the technology and its potential applications are considered from a variety of perspectives.
Simplification	Looking for innovations, people tend strongly to add new elements, making systems more complex. They tend to systematically overlook useful changes to make things simpler. Only in a small minority of cases people discover options to make things simpler.[a] With some creativity productions, systems and work organizations can be designed which comprise less (but often more versatile) units. Make it easier to do what is intended (including compliance to procedures); Make it physically impossible or difficult to do what is not intended (e.g., incorrect assembly). Make work and production processes easy to understand. Avoid overspecialization. Sometimes the credo *less is more* is valid: less detailed information means less distraction from key issues and better decision-making.[b] Making the organization simpler, often implies giving more trust and delegating responsibilities to the workers, which enables ownership, commitment and proactivity. Simplification is also relevant for communication. Many people have difficulties reading text with jargon or complex sentences. Be aware that IT has made it possible to combine many modular components of a production system into very complex production systems. A complex (work) organization with a high degree of specialization and low or no control options for the worker can be controlled centrally using IT. This is clearly counterproductive for SHW. Apart from a direct impact on health and wellbeing, the complexity of such technical or organizational arrangements implies many interfaces, many possibilities for equipment failure, loss of oversight, human error, and the possibility of *emergence* of unexpected phenomena.

(*continued*)

TABLE 6.2 (Continued)
The 12 principles for worker-friendlier design

Principles for Worker-Friendlier Design	Explanation/Example
Minimization	To minimize SHW hazards and risks in the production process is a key principle of worker-friendlier design. It ranges from eliminating hazards such as toxic chemicals and heavy loads to minimizing their use, and to minimize the division of work. Minimizing may require alternative ways of production or (re)design of work processes. When technological hazards are minimized, it usually makes maintenance simpler, safer and cheaper. The so-called socio-technical school of organizational design has clearly demonstrated that minimization of the division of labour is conducive to health and wellbeing, and is also good for safety. A low degree of division of labour and a minimum of procedures and instructions implies a relative high degree of autonomy and control for the workers. Minimization is a relevant principle for any SHW management system too. Minimize the number of procedures, so that all of the remaining paper work is strictly necessary.
Substitution	The principle of substitution is often associated with the principle of elimination or minimization. Elimination may require to substitute the most hazardous technological component by a less hazardous alternative. It can imply the substitution of hazardous tasks by technological means (e.g., robotization) or alternative ways of working e.g., the substitution of manual lifting of heavy weights by using of a lifting aid. Be attentive to new hazards and risks that can be introduced by the substitute.
Moderation	Moderation for safety refers to moderate speed, moderate height, moderate temperatures, and the use of moderate pressures. For physical burden, it may refer to moderate weights. In work processes moderation refers to a milder work load and speed, no overburdened schedules. To reasonable working times, which allow for a bearable work-life balance, etc. A good shift schedule and reasonable working times reduce the likelihood of (unnecessary) fatigue (and the consequences thereof for production, quality and SHW). Remember: to work is not a top sport, people should be able to work till their age of pensioning. With respect to conflicts and undesirable behaviour at the workplace, moderation implies having an independent confidant that has the power to promote a pleasant atmosphere and reconciliation in case of conflicts.

TABLE 6.2 (Continued)
The 12 principles for worker-friendlier design

Principles for Worker-Friendlier Design	Explanation/Example
Autonomy and control	In the design of the work organization, make sure that people have a reasonable degree of control or autonomy in their work as well as opportunities for learning and personal development. It will make the job more attractive, stimulate learning by doing and is good for wellbeing. It is enabling ownership and taking responsibility for SHW.[c] Be aware that technological choices, especially those regarding IT often implicitly determine the work organization, often with reduced autonomy and control as a result. This need not be. But it implies that autonomy and control should be integrated in the choice of technology, not thereafter.
Forgiving capability	It is important to make (technological and work) processes less vulnerable in case of failure or a human error. This may imply that some kinds of buffers or check points with an associated correction mechanism are built in. The principle is also relevant in the development of a prevention culture and how the organization deals with errors and failures. It is important to avoid blaming people and to underline the need to learn from errors. This is linked with the development of a just culture.
Optimizing the layout of production facilities and workplaces	The layout of production facilities and workplaces is relevant for the safety and health risks associated with logistics of materials, components and people. It is relevant for the location (and potential impact) of hazards in relation to the place where people work.

[a] Adams, G.S., Converse, B.A., Hales, A.H., Klotz, L.E., (2021) People systematically overlook subtractive changes. *Nature,* 592, 258–261. Doi: 10.1038/s41586-021-03380-y.

[b] Flyvbjerg, B., (2022) Heuristics for master builders: Fast and frugal ways to become a better project leader. *Saïd Business School Working Paper*, University of Oxford.

[c] See also ISO 45003, (2021) *Occupational Health and Safety Management. Psychological Health and Safety at Work. Guidelines for Managing Psychosocial Risks*. Geneva: ISO.

of office, or let's *wait for further evidence*. In this respect champions and internal sponsors for WFD can play an important stimulating role. For leaders, it is important to develop the capability to implement disrupting technologies and *out-of-the-box solutions.*[37]

Make sure to recognize and use the key moments when important choices about technology or organization are in the process of making (e.g., investment decisions, mergers and takeovers, reorganizations, determination of new product/market combinations, the adoption of new technologies). Involve key stakeholders and stimulate participation of key stakeholders including future users in experiments with new technologies and new ways of working.

An interesting example is the global framework agreement on digital transformation between the Solvay Group and Solvay Global Forum and European works Council. The agreement addresses the social aspects of digital transformation in the Solvay Group. New digital technologies will be assessed on their potential impact on SHW in the phase of project preparation and implementation. Consultation of worker representatives takes place before the decision to introduce the technology can be taken.[38]

WFD can also have implications for the investigation of events (accidents, incidents, cases of diseases, warning signals, etc.). When there are SHW issues, the focus is usually to try to get them under control as soon as possible. Then the issues are analysed to identify their causes and correct the situation. However, this leaves little time for reflection to develop more fundamental improvements, which may better prevent future reoccurrence. As Kletz already noticed:[39] 'Most systems of (accident) investigation are designed to find out why the control of hazards or risks was lost; they do not ask why the hazard was tolerated and if it could have been avoided'. One can add that mental models determine the systems we design and use, and how they are valued and analysed.[40] Mental models are therefore underlying causes of events that are usually not addressed in accident investigations.

6.5.6 Proactive Prevention during Production and Maintenance

During production and maintenance emerging problems have to be recognized and solved, while new opportunities for improvement deserve attention. Proactive prevention then refers to early anticipation of (weak) warning signals to prevent potential SHW issues, and adequate responses to incidents, near misses and health complaints (see also Section 9.4).

At any stage, it is important when problems are identified to first consider options for proactive prevention and worker-friendly (re)design before considering preventive control. In each stage, it is relevant to first get an overview of technological and organizational options, and thereby be alert to new opportunities, e.g., as a consequence of recent technological developments. Only then the best option can be chosen. Feedback loops from production and maintenance to those who were involved in development and design are important for mutual learning and continual improvement, see Figure 6.2.

The world is not ideal: zero risk does not exist. Residual risks that could not be eliminated in the early stages will have to be dealt with during production and maintenance. There are no workplaces where changes do not occur (equipment requires maintenance and repair, technology wears out over time, new software is regularly introduced and updated, team compositions change over time, new colleagues

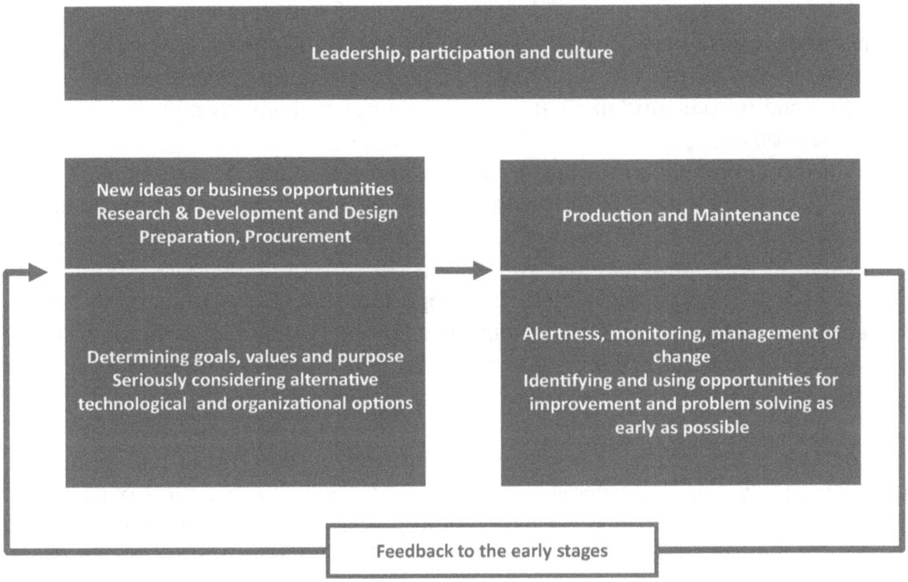

FIGURE 6.2 The feedback loop of proactive prevention from production and maintenance to the early stages of development and design.

enter the workplace, etc.). When people are more experienced and older they have to cooperate with young people who have a different type of education, etc. It is important to make sure these changes will not lead to new SHW risks and their potential consequences. Again, the earlier these risks are recognized, the more effective and cost-effective they can be tackled. The 12 principles of WFD are still relevant, but feasibility may limit their application during production and maintenance.

6.6 DIGITALIZATION AND (PROACTIVE) PREVENTION

The rapid developments in digitalization of production and work offer important opportunities for prevention, both for proactive prevention and preventive control. Through sensors and cameras, IT can *observe* situations in real time. Through its enormous calculation capacity, it can analyse data in real time, including big data sets for comparing actual information with historical information. Through Artificial Intelligence it can draw conclusions and propose decisions, and through electronic control measures it can 'do' what seems needed. In *the internet of things,* IT elements are increasingly integrated into production facilities, tools and clothing. That allows for communication and mutual interaction between the various elements of the production system. This implies that the *learning by doing* process as described by Kolb[41] is no longer only for human beings. IT can create learning processes and more intelligence in our production systems as well. Remote control makes it possible to carry out production activities at a distance, thereby significantly reducing exposure to SHW risks. Self-steering vehicles, drones and robots open up totally new possibilities for the prevention of dangerous tasks. IT is very promising for dynamic forms of preventive control: risks assessments linked to sensors, data bases and AI can take *preventive control* in real time, i.e., much earlier than ever before. The rapid evolution and introduction of IT in production implies many changes in work, jobs and professions.

IT has the potential to support *proactive prevention* too. IT can support many of the 12 principles for proactive prevention. IT can strengthen the principle of *being early.* It can support risk assessments in the early stage of design leading to the design of safer and healthier workplaces. In case of complexity, *digital twins* can be used to explore SHW risks and opportunities. IT can offer opportunities for *early anticipation* and the prevention of problems. IT can be used *to adapt the work environment and tools to the (individual) worker*, e.g., by adapting the speed of working, reducing the physical burden to the capacity of the individual, or by using remote control options to reduce exposure to dangerous conditions. The introduction of cobots makes robots much more human-friendly. IT can support the principle of *substitution,* especially by the introduction of robots or drones in case of very dangerous work (e.g., in fire-fighting) which can takeover activities from people. The introduction of self-driving vehicles is another example with the potential to improve safety. The principle of *moderation* can be supported, e.g., by moderating conditions of production (e.g., better heat control at the workplace) or by moderation of speed of production in case of a less-qualified worker. AI can *simplify* work (but also create a false sense of simplicity and safety). IT can give workers, individually but also at team level, a greater *control* over their work because IT can provide them with better and up-to-date information that would otherwise not be available. IT can strengthen the *forgiving capability* of the production system, e.g., in case of human error or technical failure. As IT can real-time identify and control errors and deviations from normal, it can support the optimization of the logistics and layout in workplaces, etc.

But IT can also introduce new risks, especially when it is not designed according to the ergonomic principle that it should support or empower people in their actions. Presently it is not exceptional that software limits workers in the *control* over their work processes. Even worse is when IT is primarily used to monitor and supervise workers and their behaviour continuously. That affects privacy and wellbeing and often also health. IT can increase power differences, while AI can unintendedly affect human rights and values, for instance, by increasing discrimination.

6.7 CONCLUSIONS

Prevention is essential for Vision Zero. However, the concept of prevention is used in very different ways. The traditional hierarchy of prevention measures has many limitations because it is not useful for the prevention of psychosocial risks, while important factors like leadership and education and training are not included. In practice, prevention is often reduced to the two lowest levels of the hierarchy: administrative controls and PPEs. These added-on prevention measures are popular because they can be implemented relatively quickly and cheaply and do not require to reconsider the way production and work processes are organized. They imply that the behaviour of the workforce becomes a key factor in prevention: the consequences are a focus on compliance to the rules, wearing PPEs and behavioural approaches with a focus on individual behaviour. In the health and wellbeing area, the distinction is made between primary, secondary and tertiary prevention. Both the hierarchy and this distinction contribute to the confusion around the concept of prevention.

All that is not helpful for Vision Zero or any other long-term strategy for prevention. That is why there is a need to redefine the concept of prevention in a proactive and strategic way, relevant to all three aspects of safety, health and wellbeing at work and Vision Zero. Therefore, the concept of *proactive prevention* was developed. It implies that the choice of technologies and organizational arrangements is relevant and includes timing, i.e., taking prevention measures as early as possible with a focus on measures integrated into the production process itself (thereby reducing the need for added-on preventive control measures). Different from proactive prevention, control measures are taken when production and work arrangements are given and added-on measures (e.g., procedures, PPE), or minor adaptations of work and production processes (e.g., job crafting) are taken to reduce the risks. Proactive prevention cannot eliminate all SHW hazards and risks, nor guarantee that no (new) SHW issues will ever arise. It remains to be alert to upcoming and emerging SHW hazards and risk and that action should be taken as early as possible.

Proactive prevention measures in the early stages will be mostly integrated into technology, work organization, work processes, tools and workplaces. It requires design thinking: generating a variety of options and then selecting the best option for the situation at hand. Twelve principles for Worker-Friendlier Design (WFD) are relevant: purpose, be early, precaution, design for people (human factors), participation, simplification, minimization, substitution, moderation, autonomy and control, forgiving capability and optimizing layout and logistics of production facilities and workplaces. While most potential SHW problems can be prevented this way, now and then new and unexpected SHW challenges will inevitably arise during production or maintenance. Again, it is important to deal with them proactively, be attentive to early warnings to anticipate the upcoming SHW problems as early as possible (see Section 9.4) and to consider technological or organizational solutions. The application of the 12 principles for WFD is also relevant in the stages of production and maintenance, though feasibility issues may lead to a preference for control measures as they require less drastic actions.

The process model for proactive prevention emphasizes not only interventions in the early stages like design, but also feedback from production and maintenance to those involved in the earlier phases to support mutual learning.

NOTES

1 ISO, (2021) *ISO Standard 22316; Security and Resilience – Vocabulary*. Geneva: International Standardisation Organisation.

2 Chambers, H., (2013) *Prevention and Control Strategies*. OSH WIKI. https://oshwiki.osha.europa.eu/en/themes/prevention-and-control-strategies. (Accessed 19 April 2021).

3 EU OSHA, (2009) *Assessment, Elimination and Substantial Reduction of Occupational Risks*. Available at: https://osha.europa.eu/en/publications/assessment-elimination-and-substantial-reduction-occupational-risks (Accessed 19 April 2021).

4 Kletz, T., (1998) *Process Plants: A Handbook for Inherently Safer Design*, Philadelphia: Taylor & Francis.
5 Hale, A., Borys, D., (2013) Working to rule, or working safely? Part 1: A state of the art review. *Safety Science*, 55, 207–221, Doi: 10.1016/j.ssci.2012.05.011; and Hale, A., Borys, D., (2013) Working to rule, or working safely? Part 2: The management of safety rules and procedures. *Safety Science*, 55, 222–231. Doi: 10.1016/j.ssci.2012.05.013.
6 Caplan, G., (1964) *Principles of Prevention Psychiatry*. Oxford, England: Basic Books.
7 E.g., Cooper, C.L., Cartwright, S., (1994). Healthy mind – healthy organization – a proactive approach to occupational stress. *Human Relations,* 47 (4), 455–471. Doi: 10.1177%2F001872679404700405.
8 Karasek, R.A., (1979) Job Demands, job decision latitude and mental strain: Implications for job redesign. *Administrative Science Quarterly*, 24, 285–308. Doi: 10.2307/2392498; and Karasek, R.A., and Theorell, T., (1990) *Healthy Work, Stress, Productivity and the reconstruction of Working Life*. New York: Basic Books.
9 ENWHP, (2018) *The Luxembourg Declaration on Workplace Health Promotion in the European Union, updated version 2018* (first version 1997): www.enwhp.org/?i=portal.en.workplace-health-promotion (Accessed 12 June 2020).
10 Le Coze, J.C., (2022) The 'new view' of human error. Origins, ambiguities, successes and critiques. *Safety Science*, 54, 105853. Doi: 10.1016/j.ssci.2022.105853.
11 Zwetsloot, G.I.J.M., Schmitt-Howe, B., Nielsen, K.T., (2020) Success factors for OSH implementation – Opening the black box of OSH Realization, *Policy and Practice in Health and Safety*. 196–210. Doi: 10.1080/14773996.2020.1786994.
12 Dyreborg, J., Johnstone Lipscomb, H., Nielsen, K., Törner, M., Rasmussen, K., Frydendall K.B., Bay, H., Gensby, U., Bengtsen E., Guldenmund, F., Kines, P., (2022) Safety interventions for the prevention of accidents at work: A systematic review. *Campbell Systematic Reviews*, e1234. Doi: 10.1002/cl2.1234.
13 Hollnagel, E., (2014) *Safety I and Safety II. The Past and Future of Safety Management*. Farnham: Ashgate.
14 Feigenbaum, A.V., (1956) Total Quality Control. *Harvard Business Review*, 34 (6), 93–101.
15 Crosby, P.B., (1979) *Quality Is Free – The Art of Making Quality Certain*. New York: McGraw-Hill.
16 Van Weenen, J.C., (1990) *Waste Prevention: Theory and Practice*. Doctoral thesis, Technical University Delft.
17 Freeman, H.M., (1995) *Industrial Pollution Prevention Handbook*. University of Michigan, McGraw-Hill.
18 Parker, S.K., Bindl, U.K., Strauss, K., (2010. Making things happen: A model of proactive motivation. *Journal of Management*, 36, 827–856. Doi: 10.1177/0149206310363732.
19 The term *proactive prevention* was recently used by several authors to address proactive (individual) behaviour aiming at prevention (e.g., through early anticipation). It was used by Njuguna et al., in the context of preventing non-communicable diseases in low-burden populations in public health policy. Njuguna, B., Fletcher, S.L., Akwanalo, C., Asante, K.P., Baumann, A., Brown, A, et al., (2020). Proactive prevention: Act now to disrupt the impending non-communicable disease crisis in low-burden populations. *PLoS ONE*, 15 (12), e0243004. Doi: 10.1371/journal.pone.0243004.
20 Note that a strong prevention culture is in line with the concept of proactive prevention as used by Njuguna et al. and some other authors.
21 Johnsen, S.O., Okstad, E., Aas, A.L., Skramstand, T., (2013) Proactive indicators of risk in remote operations of oil and gas fields. *SPE Economy & Management*, 4 (02), 90–105. Doi: 10.2118/126560-PA.
22 Kletz, T., (2001) *An Engineer's View on Human Error* (3rd edition, first edition 1985). Rugby, UK: Institution of Chemical engineers.
23 Flyvbjerg, B., (2022) Heuristics for master builders: Fast and frugal ways to become a better project leader. *Saïd Business School Working Paper*, University of Oxford.
24 The long-term impacts are also relevant for environmental reasons.
25 The lack of full scientific evidence implies decision-making under uncertainty, complexity or ambiguity. See: Stirling, A., Gee, D., (2002) Science, Precaution and Practice. *Public Health Report*, 117, 521–533.

26 EU, (2006) *Regulation Concerning the Registration, Evaluation, Authorisation and Restriction of Chemicals (REACH)*. Brussels: European Union 1907/2006.

27 Gallis, R., Zwetsloot, G.I.J.M., (2017) Paradigm shifts as root cause of accidents and disasters. Unpublished discussion paper.

28 Roggeveen, V., (2022) *The influence of leadership on the prevention of safety incidents: on risk reduction, leadership, safety principles and practices*, p. 37. Doctoral thesis. University of Leiden (NL).

29 WHO/ILO, (2021) *Joint Estimates of the Work-Related Global Burden of Disease and Injury 2006–2016 – Global Monitoring Report*. Geneva: World Health Organization and International Labour Organization. Available at: www.who.int/publications/i/item/9789240034945 (Accessed 28 February 2024).

30 Trist, E.L., Bamforth, K.W., (1951) Some social and psychological consequences of the Longwall method of coal getting. *Human Relations,* 4, 3–38.

31 Kletz, T., (1978) What you don't have can't leak. *Chemistry and Industry,* 42, 287–292.

32 Kletz, T., (1998) *Process Plants: A Handbook for Inherently Safer Design*. Philadelphia: Taylor & Francis.

33 Ashford, N.A., Zwetsloot, G.I.J.M., (2000) Encouraging inherently safer production in European firms: A report from the field, *Journal of Hazardous Materials*, 78, 123–144. Doi: 10.1016/S0304-3894(00)00220-X; Zwetsloot, G.I.J.M., Ashford, N.A., (2003) The feasibility of encouraging inherently safer oroduction in industrial firms, *Safety Science,* 41 (2), 219–240. Doi: 10.1016/S0925-7535(02)00003-6.

34 It can be broadened to include environmentally sustainable design principles.

35 Trevor Kletz introduced the term *friendlier design* for inherently safer designed process installations. Kletz, T., (1990) The need for friendly plants. *Journal of Occupational Accidents,* 13, 3–13.

36 A different but similar concept is *human-centric design*, which is regularly used in the context of industry 5.0, wherein the impact on the environment is also addressed. See Breque, M., De Nul, L., Petridis, A., (2021) Industry 5.0: Towards a sustainable, human-centric and resilient European industry. Publications Office of the European Union. Doi: 10.2777/308407

37 Ackermann, R., (2023) Design thinking was supposed to fix the world. Where did it go wrong? *MIT Technology Review*, 9 February 2023.

38 Solvay, (2020) *Global Framework Agreement on Digital Transformation between Solvay Group and Solvay Global Forum (SGF) and European Works Council (EWC)*. Brussels: Solvay.

39 Kletz, T., (1998) *Process Plants: A Handbook for Inherently Safer Design*. Philadelphia: Taylor & Francis.

40 Senge, P., (1990) *The Fifth Discipline, The Art and Practice of the Learning Organisation.* New York: Doubleday.

41 Kolb, D.A. (1984) *Experiential Learning – Experience as the Source of Learning and Development.* New Jersey: Prentice Hall.

7 Prevention Culture

One of the roots of Vision Zero, as presented in Section 3.2, was the concept of safety culture, which came to the forefront after the analysis of the Chernobyl disaster (1986). Prevention culture was already mentioned a few times in earlier chapters. A strong prevention culture is essential for Vision Zero and a vital complement to proactive prevention. In the literature, the concepts of safety culture, OSH culture and prevention culture are used as well as national-, organizational- and workplace culture. The different cultural concepts have a lot in common. In relation to Vision Zero, the term prevention culture is used on the national level[1] as well as organizational and workplace level. In this chapter, the focus is on the organizational and workplace level, whereas prevention culture at the national level is mentioned in Section 10.2 (National and international policies for Vision Zero). At each level, the fundamental right to a safe and healthy working environment is the basis for a strong prevention culture. In Section 6.5 on proactive prevention, the focus was on technical and organizational prevention measures. That is complemented in this chapter by the focus on prevention culture.

7.1 WHAT IS A PREVENTION CULTURE?

Culture refers to a group, team, department, organization, inhabitants of a region or country, always to a collective. Culture is strongly influenced by underlying factors like values (see Chapter 4) and unwritten rules that, usually unconsciously, govern collective behaviour.[2] Every organization and every group has unwritten rules. In organizations, these unwritten and informal rules can have as much impact on the behaviour of people as formal agreements or procedures;[3] in exceptional cases even more. The unwritten rules are developed through frequent social interactions and (crucial) events or developments that have taken place in the organization. The unwritten rules are transferred and confirmed by social interactions. When these rules are broken, what can easily occur in the case of new employees or outsiders, it will rapidly made clear to them what the informal rules are. The unwritten rules are often based on and influenced by informal factors which may comprise values, beliefs, habits, mental models, taboos, blind spots, as well as social interactions. The unwritten

DOI: 10.1201/9781003621225-7

rules explain why there are patterns of *cultural expressions* (often behaviours or the consequences thereof) that occur with some regularity.

The development of a prevention culture is often seen as a way to structurally improve SHW in the long range. The underlying idea is that when technology and organization are more or less optimized and under control, human behaviour, individually and collectively remains the weakest factor and can be addressed via cultural change. It is thereby assumed that subtle forms of social control can promote and ensure the desired preventive behaviour. These intangible aspects of prevention are often considered as 'soft', but they surely have 'hard' impacts. For addressing a prevention culture, it is essential to address both tangible and intangible factors as they always interact. See Table 7.1.

As described in Chapter 2, we live in a world of change. Organizations tend to become increasingly fragmented and temporary. Fragmented in terms of space by distant work, in terms of social interactions through outsourcing of activities and working with temporary personnel, contractors and subcontractors, and involving self-employed persons. This weakens organizations as social communities. As a result, companies can become somewhat reluctant to invest in their people while the personnel that is working for them might be less interested in contributing to the (strategic) aims of the organizations they temporarily work for. These developments undermine the formal prevention arrangements. They increase the importance of the intangible aspects such as culture and values, as they remain relatively stable over time (except when partnerships or the composition of personnel change very quickly and drastically). That has positive and negative consequences. Positive is that the culture can be a stabilizing factor in times of drastic changes. Negative is that culture is difficult to change. The development of a strong prevention culture therefore requires a long-term strategy, visible leadership commitment and perseverance.

TABLE 7.1
Examples of tangible and intangible factors that codetermine prevention culture

Tangible Factors	*Intangible Factors*
Strategy	Mental models
Structure	Culture, values
Rational decision-making	Taboos, blind spots, tunnel visions, hidden agendas, emotions
Risk assessment	Risk perceptions
Hierarchical power	The power of local heroes
Instructions	Exemplary behaviour
The employment contract	The psychological contract
Written procedures, work as imagined	Practices in dealing with procedures, work as done
Written and verbal messages	Used language, tone of voice, body language
Qualification, diplomas	Experience and tacit knowledge

The concept of prevention culture is a multifaceted and multi-layered concept. As the importance of facets and layers may depend on the context, this makes it difficult to harmonize definitions and measurements. For the definition of a prevention culture, the following four issues are key.[4]

1. The interactions of a prevention culture with SHW management (systems or policies) are important. Prevention culture and SHW management are mutually dependent. Both are necessary for excellent SHW performance.
2. The organizational prevention culture is part of the broader organizational (and societal) culture. While the prevention culture is focused on SHW opportunities and risks, it is strongly influenced by broader cultural factors. Examples are trust in the organization or individualism in society.[5]
3. For SHW, the practices in organizations are of utmost importance for SHW performance. But the practices are influenced by deeper layers, such as values, beliefs and mental models. These are important underlying factors of a prevention culture and are crucial for the long-term development of a strong prevention culture.
4. The concept of prevention culture in itself is neutral. A prevention culture can be strong or weak, proactive or reactive. Developing and improving the prevention culture is a challenge. A culture is always dynamic and its gradual development can be positive or negative.

Here we define prevention culture as: the shared attitudes, values, mental models, perceptions and habits of the majority of the members of an organization (or organizational unit), relevant to dealing with SHW opportunities and risks.[6] In simple words: a prevention culture is about how the majority of the people take care of each other, their work environment, the work processes, etc.

Prevention culture is a complex issue that cannot be directly observed or measured. Only the expressions of a culture can be directly observed. It is important to realize that one single event, e.g., the behaviour of an individual that has positive or negative consequences for SHW could be an expression of the culture. One observation is, however, never sufficient to formulate conclusions on the quality of the prevention culture. Several coherent observations are needed to come to such conclusions. The informal side of the organization is almost by definition less transparent than the formal side; for a significant part, it is subconscious or unconscious for the people involved. Surveys cannot measure the unconscious aspects of a prevention culture; surveys therefore have important limitations in this respect.

7.2 FORMAL AND INFORMAL PREVENTION

Culture, the intangible side of the organization or organizational unit, is a necessary complement of formal prevention with its technical and organizational measures, its rules and prescriptions. Preventive activities without a strong prevention culture may lead to compliance but will not generate SHW excellence. On the other hand, developing a prevention culture without taking formal and tangible preventive measures will not be credible and not be successful either. People need two legs to take steps forward; similarly, organizations need *two legs* for organizational excellence in SHW: the formal and the informal part of the organization have to work together. The development of a strong prevention culture and SHW management should go hand in hand in achieving high levels of prevention: both *legs* are required to make progress.

Some people believe that prevention depends only on the formal part as in technical and organizational prevention measures that are consciously and rationally taken. However, that leads easily to a focus on procedural compliance to legislative requirements. It may result in a technological fix or on SHW prevention being well-organized on paper, but with limited impact on SHW performance. Unfortunately, unintended consequences of formal preventive arrangements are not exceptional. Too many rules and prescriptions reduce the possibility for adaptation and innovation and tend to direct the culture on rule-based compliance. This is especially the case when informal social aspects are not considered and workers are not involved in the development and implementation of prevention arrangements.

For prevention, it is important to focus clearly on the practices. Procedures may look nice and useful on paper but in reality, the work might be carried out somewhat differently. That is the well-known gap between Work as Imagined (WaI) and Work as Done (WaD).[7] Something similar may be the case at hand concerning the prevention culture: often there are officially espoused values and messages concerning behaviours and practices relevant for prevention. That is Culture as Espoused (CaE). But the reality might be different and influenced by unclear social mechanisms. That is Culture as Lived (CaL). While the focus on WaD is on the behaviour of individual workers, Culture as Lived focuses on 'how we do things around here', i.e., on the practices at group or organizational level rather than that of individuals.

Formal prevention measures are usually the result of conscious decision-making and it seems such measures can be implemented relatively quickly. But formal prevention may unintendedly hinder the development of a proactive prevention culture.[8] E.g., too much paperwork and bureaucracy will reduce the more intangible aspects such as learning on the job, intrinsic motivation and autonomy. Strict managerial supervision may be perceived as distrust by the workers. Detailed protocols and bombardment of safety instructions may make workers feel disrespected or distrusted in their professionalism. This may trigger workers' perception that the protection of the organization against liabilities is more important than the protection of its workers.

The impact of the prevention culture on formal prevention can be positive or negative, depending on the quality of the prevention culture and the development level of the formal prevention arrangements. See Table 7.2. Prevention culture can therefore impact the SHW performance level positively as well as negatively.

Usually, the focus in addressing prevention and culture is on the 'sharp end', i.e., on the operational level. But the focus should also be put on the strategic level, where key decisions are taken. In most literatures, the latter is addressed as a leadership issue, not a cultural issue. However, the behaviour and decisions of the leaders are not only influencing the culture, but they are to some extent also an expression thereof. When all goes well, the formal and informal factors will strengthen each other: the

TABLE 7.2
The impacts of prevention culture on formal prevention

A poor prevention culture ...	A strong and proactive prevention culture ....
Regularly overlooks opportunities for SHW improvement	Enables the identification and use of opportunities for SHW improvement
Weakens the (shared) awareness on SHW risks	Strengthens the (shared) awareness of SHW risks
Weakens the practices related to formal procedures and decisions (e.g., those documented in the SHW management system)	Strengthens the practices related to formal procedures and decisions (e.g., those documented in the SHW management system)
Is likely to neglect and underuse opportunities for SHW improvement	Is conducive to identifying and realizing opportunities for improvement
Informally (and often unconsciously) tolerates risky SHW situations to persist; The handling thereof may suffer from unnecessary delays	Does not tolerate dangerous or risky SHW situations; these will be handled well and timely
Tolerates – at least for some time – unacceptable situations	Enables the elimination of unacceptable situations
Tolerates that available SHW information is of poor quality	Ensures that reliable SHW information is always available
Frustrates effective communication	Enables that important information is proactively shared
Leads to too much troubleshooting (unplanned maintenance and repair)	Enables preventive maintenance
Tolerates unsafe or unhealthy behaviour	Enables and promotes safe and healthy behaviour at all times
Undermines safety margins for SHW	Increases safety margins for SHW
Suggests that SHW does not require further efforts	Makes SHW a continuous challenge and learning process
Overlooks or neglects early warnings	Enables the identification of early warnings

prevention culture will then strengthen the functioning and effectiveness of the SHW policy and management, and vice versa. Formal and informal factors can reinforce each other but may also contradict each other. For example, 'the local hero' in a team can sometimes have more influence on the behaviour of the co-workers than the manager. The informal factors can indeed frustrate the proper functioning of SHW policies and management. A strong and proactive prevention culture will strengthen formal (proactive) prevention while a poor prevention culture will weaken formal prevention. When the focus is only on the formal prevention, it is likely that SHW performance will plateau and attempts to further improve SHW performance have little or no impact. In order to improve prevention, it is then necessary to recognize the role of culture.

A strong prevention culture promotes collective SHW-conducive behaviour in all phases of the production lifecycle. It generates the conditions that attention to SHW goes without saying from the early stages of research, development and design to production and maintenance.

7.3 A STRONG PREVENTION CULTURE

Several authors have mentioned key characteristics of a strong safety culture and more recently of a broad prevention culture in organizations.[9] As each organization is unique and organizational cultures are unique, these characteristics have a rather abstract level. Operationalization for the organizational context, values and practices is needed. The main characteristics of a strong prevention culture and examples of operationalization into organizational practice are given in Table 7.3.

Learning organizations explicitly manage their learning processes so that they learn to improve their learning. In practice that will often imply stimulating dialogues, offering opportunities to reflect on mental models and strengthening and promoting the active participation of internal as well as external stakeholders.

Two closely related value aspects that are conducive to a learning prevention culture are trust and the absence of fear and anxiety. Trust and the absence of fear and anxiety are important for people at the shopfloor to feel free to report incidents and near misses, as well as to come up with suggestions for improvement.[10] In the absence of fear, errors can be freely discussed and reported to learn from. The absence of anxiety is, however, also important in the higher echelons of the organization. When higher managers are afraid that they might not meet their targets, they may neglect information about possible risks, as well as indications of emerging dangers. When notifications from people are neglected or are heard without leading to any consequences, people will stop reporting. As a consequence, the higher managers remain deprived of critical information. That strengthens the idea that everything goes well, which easily results in the underestimation of SHW risks.

The benefits of a strong prevention culture are: fewer SHW problems and associated costs and more productive employees; fewer quality and delivery issues, higher reliability, flexibility and productivity in production and work processes; a better status in society and attractiveness in the labour market, greater profitability and several social and economic benefits for society at large. When a prevention culture is strongly focusing on higher-order learning, the cultural approach will help to realize important (SHW) innovations that go beyond merely optimizing the existing socio-technical system. The focus on learning, participation, by well-informed people implies that a strong prevention culture supports commitment strategies and increases the social capital of the organization. Social capital refers to the resources and (potential) benefits associated with social relationships, such as having networks, sharing norms and trust, which facilitates actions by the individual as well as the collective. Social capital is important for organizational development and successful management of change, innovation and SHW.[11]

Several cultural factors can only be promoted and enabled. See Table 7.4. Note that each of these factors is good for SHW and for professional development.

TABLE 7.3
Characteristics of a strong prevention culture

Characteristics of a strong prevention culture	Operationalization: Examples of practical consequences
SHW is adopted as a core value; It is integrated in each process and workplace	SHW is seriously considered in each important business decision, unnecessary risks are thereby avoided
	No activity is important enough to neglect SHW
	SHW of contractors and third parties is just as important as SHW of own personnel
	People contribute actively to SHW because they want it, not because they have to.
There is clear SHW leadership	Leaders are personally committed to SHW; it is *in their DNA*
	Leaders' verbal messages and behaviour are aligned
	Leaders take responsibility for the underlying causes of SHW problems
	Concerns or ideas regarding SHW can always be discussed at any time, regardless of hierarchy or position
	Workers know they will be supported by their leaders in case of SHW difficulties and challenges.
	Leaders regularly visit all departments to learn about the risks, challenges and opportunities workers experience in their work
Accountability for SHW is clear	When there is an acute danger, work will (always) be suspended
	Everybody knows at all times who is responsible for what
	When a SHW problem is identified, action will be taken shortly (otherwise it becomes a chronic problem that people tend to tolerate)
SHW is learning driven	People are willing to reflect on their own and other's behaviour
	Adaptation of work and processes to new circumstances is normal practice
	There is the shared awareness that there is always room for further improvement
	It is not regarded problematic that people err, as long as errors can be openly discussed and are used for learning
	Critical positive feedback is normal practice
	It is encouraged to experiment, within certain conditions
The value of good jobs is recognized	The work organization implies that people have a reasonable control over their work processes
	The work organization enables learning by doing and personal growth
Participation	Workers are expected and encouraged to actively and critically reflect on their work; their ideas are taken seriously
	Workers feel free to come up with suggestions for improvement or to report problems
	People feel responsible for the SHW of others

TABLE 7.3 (Continued)
Characteristics of a strong prevention culture

Characteristics of a strong prevention culture	Operationalization: Examples of practical consequences
Everybody is well-informed	Managers have knowledge about the human, technical, organizational and environmental factors that determine SHW Workers have all information that is needed to work safely and healthy All paper work is meaningful and up to date At the start of activities, the SHW risks are always assessed and discussed
Reporting is encouraged	Dangerous situations are recognized, discussed and reported People are willing and encouraged to report near misses, warning signals, errors and complaints Workers who address or report SHW issues receive adequate and fair feedback in a timely manner It is good for making career to report SHW issues or errors
Justice	Fair processes in selection, evaluation and promotion of personnel Providing reliable information is rewarded, even if it is 'bad news' Inappropriate behaviour is not tolerated; when it happens, a serious conversation will follow, and if necessary, measures will be taken to prevent reoccurrence
A supportive culture	Managers and workers give and receive active social support Social support both by leaders and colleagues is important for psychological safety. That is necessary to speak up, to discuss sensitive issues and to report incidents, complaints, etc.
Align and balance work and life	Time for recovery and private life are as important as working hours Responsible behaviour in private time is encouraged It might be necessary to get rid of (or limit) cultural norms and organizational habits that neglect the importance of time for recovery and private life (e.g., working excessively long hours, to feel guilty when leaving on time, to be 24/7 open to receive and respond to messages)

TABLE 7.4
Examples of cultural factors that cannot be achieved by rational planning or external incentives

Adoption of human and social values, alertness, creativity, open and honest communication, commitment, delayed judgement, higher-order learning, intrinsic motivation, style of leadership, mental models, ownership, personal commitment, proactive behaviour, taking care, taking responsibility, trust

7.4 THE IMPORTANCE OF COMMUNICATION

Vision Zero provides direction for companies in a process of continual improvement of SHW performance. Initiatives or programs are provided to tackle existing SHW problems and challenges and to realize improvements. It is important that activities and successes are broadly communicated (e.g., to other departments) and experienced on the shop floor. Otherwise, the adoption of Vision Zero would lose credibility in the eyes of the workforce.

7.4.1 Social Marketing of SHW – Sending the Vision Zero Message

When you analyse advertisements for Nike, it is clear that they do not emphasize the quality of their shoes or T-shirts, but the identity of the users. That is apparently more convincing for potential clients than the quality of their products. The intention is that potential clients will identify themselves with the well-known influencers. Coca Cola does something similar; they communicate youthful freshness and enthusiasm instead of the taste of their drinks. Their brand becomes more attractive this way. These are forms of symbolic communication: The communication is not about a specific quality of their products, but about the identity of their potential clients. To some extent, Vision Zero can be a topic for symbolic communication as well: people and organizations want to join the Vision Zero movement to improve SHW; every accident or case of work-related ill health is one to many; improvement is always possible, SHW always matters!

Paying attention to many minor problems that do not pose a serious threat to SHW can be a useful strategy to involve people, to make them aware that improvement is possible and to celebrate the first successes. In that strategy, again, the symbolic meaning is more important than the minor actual progress. To be credible, it is then, of course, essential not to neglect more serious issues. In case it requires more time to solve the serious issues, it is important to communicate that it is work in progress and that the problems will be solved within a somewhat longer timeframe.

Another example of symbolic communication is the organization of a Vision Zero day, wherein SHW is put central, not production, and all members of the organization are invited to suggest ideas for SHW improvements. Apart from the program and the usefulness of the suggestions received, to organize such a day sends a clear symbolic message to the internal and external stakeholders: We invest in Vision Zero; SHW prevention really matters, even when it implies reduced production.

Feedback and reflection on how messages are perceived by the 'target group' is important. Is everything that is communicated officially regarded as 'credible' by the workforce? Or rather as a form of public relations or even as blah blah? The credibility of the message will to a great extent depend on the credibility of the sender, i.e., how genuinely the leaders are committed to SHW. If that commitment is only a word of mouth, the messages will be received with scepticism and have little impact. The credibility of management and of SHW messages can be undermined when the use of power is stronger than rational argumentation when mistakes of management are covered up while those at the shop floor are highlighted, or when the experiences of stakeholders are neglected.

Well-known deficiencies in communication are: sharing information that is incomplete, wrongly assume sufficient prior knowledge, using a language that is too difficult for the target group, neglecting the context or reframing information to make the SHW situation seem better than it actually is.

7.4.2 Two-Way Communication

Good communication is a key factor in building a strong prevention culture. It is not only a matter of sending the right messages. It is perhaps even more a matter of good listening. Communication is reciprocal and often requires mutual feedback. Dialogues are therefore the number one tool for good communication.

Dialogues are important around accidents, incidents and cases of work-related ill health. Periodic evaluations – of what has happened in the last few months – and meetings to discuss with the relevant stakeholders what lessons can be learned from the experiences are very useful. This should not be limited to problems encountered: it is also necessary to learn from activities that went unexpectedly well.

Participation implies formal and informal two-way communication and stimulates ownership. Ownership is conducive to internal motivation, proactivity and other forms of desired behaviour that cannot be generated by coercion or economic incentives.

7.4.3 The Importance of Feedback

When people are encouraged to come up with suggestions for improvement, these suggestions and the individuals involved need to be taken seriously. Suggestions need to be evaluated, fairly and promptly. The reporter should be thanked for the suggestion and needs to receive adequate and honest feedback, also in case the suggestion turns out not to be useful. Communication should be frank and open, without hidden agendas.

Open communication can be frustrated by blame. When the messenger is held accountable for bad news, when there is always somebody who gets the blame, it becomes difficult to get all relevant information on the table. The same could happen when in practice other issues turn out to be more important than SHW (e.g., risk-taking is good for your career).

People with different cultural backgrounds, values or mother tongues may understand communication or feedback in different ways. Training interculture communication and cooperation, dialogue and sharing knowledge may overcome these kinds of communication problems. Feedback is more than solely the contents of a message; it also has a social and emotional dimension.

An important tool for communication is the management walk-around. The aim is not to inspect the workplace. For managers, it is an opportunity to learn firsthand from the experiences and challenges experienced on the shop floor, and at the same time show personal commitment to SHW; for workers, it is a scarce opportunity to communicate directly with people from higher echelons. It is best to initiate dialogues with operational people with a greater emphasis on listening and learning than on sending messages. See also Chapter 5 on leadership. A proactive culture is characterized by proactive feedback-seeking, active dialogues on improvements of work methods and proactive problem-solving.

7.5 BARRIERS TO A STRONG PREVENTION CULTURE

An organizational culture usually includes taboos and blind spots which may easily imply that some specific serious SHW issues are overlooked and not addressed. For SHW improvement it is important to recognize and fight taboos and blind spots, maybe even more than contributing to the further development of already well-developed aspects of the prevention culture.

Within a culture **taboos** are those issues that are regarded as forbidden to do, to discuss or mention or even to think about. As in 'normal life', taboos are based on unwritten rules that everybody knows, but that are seldom or never discussed. Taboos are not rational and always sensitive. Addressing a taboo violates common values and triggers emotions: people become angry or anxious; it will often lead to conflicts or social sanctions, see Table 7.5.

To recognize taboos is an important step. But then it is still difficult to eliminate them. The difficulty lies in the underlying normative aspects which people already have collectively internalized. A well-known example is transgressive behaviour in sectors like arts, media and fashion. When unacceptable behaviour is not uncommon and is more or less known to everybody, it is easily justified and accepted as 'part of how we do things around here' (in this profession, this organization, etc.). Taboos cannot be neutralized purely by rational explanations. Story-telling may be more effective than rational argumentation. Dialogues with all stakeholders involved are of help.

Blind spots are issues or circumstances that are systematically overlooked, that people are not aware of and that are not recognized. Blind spots are unconscious. Blind spots usually result from mental models and imply a way to perceive things. A blind spot can be associated with a tunnel vision, i.e., putting relevant factors aside because they are regarded as non-relevant. Different from taboos, bind spots are not necessarily value-laden; they are usually easier to discuss without triggering emotions. See Table 7.6.

TABLE 7.5
Some examples of taboos that can affect SHW

Some examples of taboos in the context of prevention culture
• Discussing the role of higher managers in the creation or continuation of serious risks • Correcting the behaviour of someone higher in the hierarchy • Addressing risks in a situation that is regarded as safe and sound • Addressing a colleague who is doing something unsafe or unhealthy; that is regarded as snitching your mates • It is preferred not to know about certain issues; when addressed they are quickly covered up • The existence of hidden agendas

TABLE 7.6
Some examples of blind spots

- In an organization with mostly technological-oriented people, problems, their causes and possible solutions are framed as technological issues (neglecting, e.g., behavioural or communication factors)
- The organization acknowledges good practices in the own sector but is not aware of or interested in those from other sectors
- Management is unaware of and not interested in the tacit knowledge of the people on the shop floor
- Management is unaware that they do not receive important critical information

To identify blind spots, outsiders are usually indispensable. Blind spots are easier identified by external consultants, inspectors, external auditors or visitors. New personnel can play a positive role as well. People within the organization are likely to share their blind spots though it may vary per department or unit. This is a problem in case of self-assessments or internal audits: then the assessors are usually not able to recognize the blind spots of their own organization.

INVITING VISITORS TO IDENTIFY BLIND SPOTS

Many companies prefer to impress visitors with some nice words of welcome by a senior manager, exciting presentations, a short tour and some snacks and drinks. But these companies might miss an important opportunity. Visitors are generally best in noticing peculiarities and remarkable practices. These can be inspiring practices the visitor can learn from. At other moments or places, they may think to themselves: *why are they doing this? This can be done so much better*! The latter implies opportunities to learn from, and sometimes address a blind spot of the receiving organization. Don't waste such opportunities! (National or sector) culture may play a role here. In some cultures, providing feedback for improvement is more sensitive than in other cultures.[12] Openness to (constructive) criticism and feedback can be an important facilitator for organizational learning.

In case your company receives visitors, tell them in advance that at the end of the visit, the visitors will be invited to give feedback. Ask them explicitly what was great and interesting, and what could be improved. It is best to give them a small piece of paper and ask them to give their feedback in writing – to communicate suggestions for improvement verbally can be perceived as unpolite.

The explicit invitation to give feedback at the end of the visit, including suggestions for improvement, is a characteristic of a learning-driven culture. It may help to overcome potential hesitations to give critical feedback or suggestions from the side of the visitors. With a view on potential blind spots of the organization, feedback from visitors should be treated just as serious as suggestions for improvement from the own workforce.

Eliminating blind spots is easier than eliminating taboos, as they are less normative. But the associated habits may still have normative aspects. A mix of storytelling and rational argumentation is adequate in most cases. When the blind spot has led to accidents or a burn-out, it can be useful to involve victims to tell the story from their point of view. That requires good preparation and social support from the organization. Mental models can also be a barrier to proactive prevention and a strong prevention culture. Some examples are presented in Table 7.7

Mental models that hinder the development of a strong prevention culture may be associated with a blind spot, but not necessarily so. While a blind spot is unconscious, a mental model is conscious and can even be regarded as rational.

Economic factors can form a fourth type of barrier to a strong prevention culture. Most important are perverse incentives and limited economic reasoning, often associated with certain mental models. Some examples thereof are presented in Table 7.8.

TABLE 7.7
Examples of mental models as barriers for (the development of) strong prevention culture

- Lack of understanding of the importance of prevention in the early stages
- Regarding prevention as a cost factor only
- Limited knowledge about the existence of alternative technological or organizational options
- Lack of experience with alternative technologies or alternative organizational arrangements
- The comfort of the existing situation versus the challenge of innovation and change
- Neglect of weak signals as irrelevant
- Seeing health issues as no more than personal individual challenges

TABLE 7.8
Economic barriers for (the development of) a strong prevention culture

- Perverse incentives
- Shifting costs of ill health to families and society
- Short term focus
- Productivity first, at all times
- Preference for proven technologies; costs of patents
- Standardization, e.g., between several plants of a corporation
- Dependence on business partners

7.6 STAGES OF DEVELOPMENT

Both formal prevention arrangements and prevention culture have several stages of development. This is well-known for safety culture, but often overlooked for formal prevention. It is important to distinguish between the development stages of the formal and the cultural (informal) prevention. An imbalance often impedes further progress. Many companies focus on compliance for their formal prevention activities but want to achieve a proactive safety or SHW culture. That is an internal contradiction. The development of a proactive culture can only be successful when it goes together with the development of proactive formal prevention activities. When the organization focuses on compliance, pushing a proactive culture would imply a disbalance between the responsibilities of management and shop floor. This is nicely expressed by Joseph Juran: '*If management tries to motivate the workers to do a better job without first cleaning up its own house, the tendency is for the people to see them coming with dirty hands*'.[13]

Usually, five stages of development or maturity are distinguished. Here we distinguish five different stages that are relevant for the formal as well as the cultural side of prevention: passive, reactive, compliance-oriented, proactive and synergetic. See Figure 7.1.

The crucial step for Vision Zero is the transformation of a culture from *we have to* into *we want to*. This is the transition from a culture focused on compliance (level 3) to a culture of proactivity (level 4). This is a much more fundamental step than the other development steps, as the motivation and the associated mental models are fundamentally different.

Until level 3 (compliance), it is possible to rely on external motivation, i.e., on orders, rules and prescriptions combined with supervision. For levels 4 and 5 (proactive and synergetic) that is no longer sufficient, even counterproductive. Ownership and intrinsic motivation are then essential. This implies that the informal aspects of

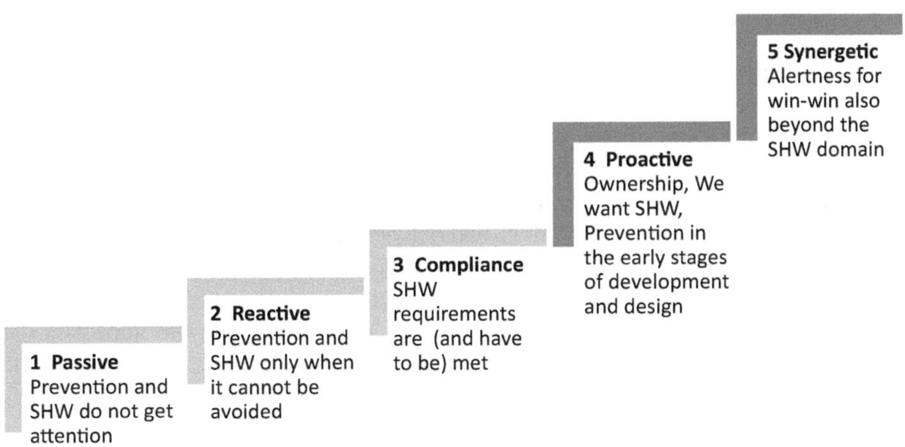

FIGURE 7.1 The five development stages of a prevention culture and of formal prevention.

the culture become relatively more important compared to the earlier stages wherein formal policies and prevention activities are dominant. For levels 4 and 5, higher-order (innovative) learning processes are crucial – and indeed, that requires ownership and intrinsic motivation from the top to the shopfloor. As a consequence, only top-down steering will not be sufficient.

The transformation from compliance to the proactive level usually starts with a shift in mindset at the top of the organization. But it becomes an organizational transformation only when that shift is also made at the operational level. That will take a process of several years.

The synergetic nature of a strong prevention culture is often overlooked (level 5). It does not only require intrinsic motivation but also an open mind, and time to reflect and consider options beyond the traditional SHW domain. It implies that SHW is no longer addressed in three different silos, and that potential positive interactions with business processes and aspects like quality, sustainability, process safety, security, etc., are consciously addressed in a learning and flexible way. That enlarges the 'system view' on SHW and implies more opportunities for prevention.[14]

As SHW is still often organized in three different silos and the three aspects have different histories, it is often found that one of the three SHW aspects is further developed than the other two. In technology-oriented industries and in the construction sector, safety is usually further developed than health and wellbeing; in office environments, there is usually more focus on health and wellbeing.

External stakeholders (policymakers, inspectors, auditors, customers, neighbours, etc.) can have an interest to recognize and promote a proactive and synergetic organizational prevention culture as an enabler of good SHW performance. For them, it is easier to observe some cultural characteristics like attitudes and ways of communication than to understand the qualities or deficiencies of the technical aspects of prevention. Outsiders or non-specialists may therefore be disproportionately interested in prevention culture.

7.7 CONCLUSIONS

Prevention culture refers to the shared values, attitudes, mental models, perceptions and habits of the majority of the members of an organization relevant to dealing with SHW opportunities and risks. It is governed by intangible factors like unwritten rules, group norms and social interactions.

Prevention culture can be strong or weak, proactive or reactive. Intangible factors, like exemplary behaviour, tone of voice and risk perceptions play a major role. Culture is difficult to influence; the development of a strong prevention usually requires at least five years; but it can also be a stabilizing factor in times of change.

Prevention culture and formal prevention are interdependent. The impact of the prevention culture on formal prevention can be positive or negative, depending on the quality of the prevention culture.

The focus of prevention culture should not be only on the operational level, as the strategic level where key decisions are taken is crucial. Leadership is essential to influence not only formal prevention but also the intangible aspects of a prevention culture. Regular management walkarounds are important opportunities to demonstrate the commitments of senior leaders.

Key characteristics of a strong prevention culture are that prevention is integrated into the core values and strategy of the organization; and that the culture is learning-driven, proactive and synergetic. These are abstract terms that have to get meaning for the practices in the organization. This can best be developed through dialogues. Trust and the absence of fear and anxiety are vital. Fear impedes discussion on sensitive issues and reporting of errors, incidents and surprises. A strong prevention culture impacts not only the stage of production but also the earlier stages of development and design.

Any organizational culture includes some taboos and blind spots that may hinder the development of a strong prevention culture. Taboos and blind spots are both unconscious. Addressing a taboo triggers emotion; blind spots are less normative, but external eyes are often indispensable for both. Mental models and economic pressure may also hinder the development of a strong prevention culture.

Five stages can be distinguished in the development of a prevention culture, as well as in the formal prevention arrangements: passive, reactive, compliance, proactive and synergetic. It is important to realize that the step from compliance to proactive is a fundamental one: it implies a turning point for the development of the organization and its members: from *we have to* (external motivation) into *we want to* (intrinsic motivation). The step to a synergetic culture implies that 'the system' considered is larger than SHW. That implies more options for prevention.

NOTES

1 ILO, (2008) *Seoul Declaration on Safety and Health at Work.* Seoul: ILO, ISSA and KOSHA. Available at www.ilo.org/resource/seoul-declaration-safety-and-health-work (Accessed 22 December 2024).

2 Guldenmund, F.W., (2000) The nature of safety culture: a review of theory and research. *Safety Science,* 34 (1–3), 215–217. Doi: 10.1016/S0925-7535(00)00014-X.

3 Scott Morgan, P., (1994) *The Unwritten Rules of the Game.* New York: McGraw Hill.

4 Zwetsloot, G.I.J.M., Van Middelaar, J., Van der Beek, D., (2020) Repeated assessment of process safety culture in major hazard industries in the Rotterdam Region (Netherlands). *Journal of Cleaner Production,* 257. Doi: 10.1016/j.jclepro.2020.120540.

5 A strong organizational prevention culture is not only focused internally but also on dealing with societal developments that may (negatively or positively) impact 'how we do things around here'.

6 Zwetsloot, G.I.J.M., Dijkman, A., (2010) *Werken aan veiligheids- en gezondheidscultuur (Working towards a Health & Safety Culture)*, The Hague: SDU.

7 Hollnagel, E., (2015) Why Is Work-as-Imagined different from Work-as-Done? In: Wears, R.L., Hollnagel, E., Braithwaite, J., *Resilient Health Care, Volume 2.* CRC Press. In earlier publications of Chris Argyris and Donald Schön on organizational learning this was called 'espoused theory' versus 'theory in use'; Argyris, C., Schön, D.A., (1979) *Organizational Learning: A Theory of Action Perspective.* Reading: Addison Wesley.

8 Edwards, M., Jabs, L.B., (2009) When safety culture backfires: unintended consequences of half-shared governance in a high-tech workplace, *Social Science Journal*, 46, 707–723. Doi: 10.1016/j.soscij.2009.05.007edwards.

9 INSAG, (1986) *Summary Report on the Post-Accident Review on the Chernobyl Accident.* Safety Series No. 75, International Nuclear Safety Advisory Group. Vienna: IAEA; Reason, J., (1997) *Managing the Risks of Organizational Accidents.* Farnham: Ashgate. Schulte, P.A., Guerin, R.J., Okun, A.H. (2014) Prevention culture as culture: Can we achieve it, and is it enough? In: *From Risks*

to Vision Zero, Helsinki 2014. Proceedings of the International Symposium on Culture of Prevention – Future Approaches. Helsinki: FIOH. pp. 21–29.

10 Edmonson, A.C., (2018) *The Fearless Organization.* Hoboken, NJ: Wiley.

11 Oksanen, T., Kouvonen, A, Kivimäki, M., et al., (2008) Social capital at work as a predictor of employee health: Multilevel evidence form work units in Finland. *Social Science and Medicine*, 66, 637–649. Doi: 10.1016/j.socscimed.2007.10.013.

12 Interesting examples on how feedback in international settings can lead to misunderstandings are given by Erin Meyer in her book *The Culture Map*. Meyer, E., (2014) *The Culture Map: Breaking through the Invisible Boundaries of Global Business.* New York: Public Affairs.

13 Juan, J.M., (1988) *Juran on Planning for Quality*. New York, London: Free Press.

14 Compare with Asby's law of requisite variety. Ashby, W.R., (1956) Self-regulation and requisite variety, in: Ashby, W.R., *Introduction to Cybernetics.* New York: Wiley. Simply said: The larger the variety of actions available to a system, the larger the variety of options for prevention.

8 Strategies for Long-Term Continual Improvement

The strategy of Vision Zero is to manage long-term processes of continual improvement in order to reduce or eliminate SHW problems, use opportunities for improvement and achieve excellence in SHW. This requires a long-term perspective. It is regularly assumed that every improvement is a step in a stepwise process of continual improvement. Unfortunately, that assumption is often wrong; short-term improvements may hinder more far going long-term improvements and can even be counterproductive from that perspective. In the business world, it is easy to see that the short-term perspective of most shareholders hinders the development of business sustainability; governments are usually weak in addressing long-term problems because politicians are focussed on the period till the next election data. Short-term solutions usually do not address causal factors at the system level. The focus is often on fighting symptoms and removing direct causes. That offers only partial solutions. Short-term solutions do not require a serious reflection or evaluation of the characteristics of the prevailing (socio-technical) system or of the existing mental models. Short-term success can, however, be important as part of a long-term strategy. It is thereby essential that the small steps are put in a long-term perspective like Vision Zero, especially when they are associated with shared values.[1] Small steps can contribute to a prevention culture wherein there is a continual search for SHW improvement.[2]

In this chapter, the focus is on optimization and innovation as the two overarching strategies for long-term continual improvement and a variety of supporting strategies. First, these two main strategies are introduced concisely. Then the supporting processes will be presented and discussed. The chapter ends with a more in-depth discussion of the two long-term strategies, their similarities, differences and complementarities, their relationship with the supporting processes and their relevance for Vision Zero.

8.1 OPTIMIZATION AND INNOVATION

8.1.1 Optimization

Optimization is a broadly accepted strategy to improve production and work processes and the impact of those processes on profits, SHW, quality, the environment and so

DOI: 10.1201/9781003621225-8

on. It implies that existing and potential problems are successively solved, and inefficiencies such as unnecessary variabilities and ambiguities are reduced or eliminated. Optimization can be done in a stepwise process whereby each step contributes to improvement. It does not require dramatic changes in the way production is organized but improves the existing system. That makes it an attractive strategy. The aim of SHW optimization in organizations is to optimize the existing socio-technical system. That will increase productivity and profitability while disturbances and undesirable impacts such as accidents, ill-health, pollution and quality problems are minimized. A well-known example is lean production. A limitation of optimization is that it takes the existing situation, i.e., the existing socio-technical system with its aims, technologies and organization, as given. By solving technological and organizational problems, the production and work processes will flow more smoothly and safe, but the hazards and risks that are inherent to the technology and the organization will remain present, though most risks are largely under control. If you optimize a push bike, you will get a perfect push bike, but you will never get an electric bike.

8.1.1.1 The Balance between Perfection and Good Enough

An often-forgotten aspect of optimization is to achieve the balance between *good enough* and *perfection*. The case of an engineering firm doing maintenance and repair work on technical installations is a nice example. The higher managers tried to convince the engineers that *good enough* is good enough. When the technological functions used by the client are OK, the product is according to the managers good enough. The engineers had higher professional standards. They felt responsible also for less vital functions of the technology. Perfection sounds perfect, but striving for perfection often works out negatively. For organizations, it might be too expensive. In complex systems it is impossible to foresee everything, implying that in such cases perfection is simply impossible. For people, perfection contributes to stress and may lead to burnout. So, when is it good *enough?* When something is *good enough* it is not perfect. It will have some shortcomings. When people work *good enough*, their work may result in some deficiencies or undesirable side effects. Nevertheless, good enough is important for work, for products and for life. It is a leading principle for policies and procedures. It is why people always say that improvement is still possible. At the same time, *good enough* is contributing to complexity and surprises.

The balance between perfection and good enough is directly related to the issue of control and its limitations. There are many examples where the pressure to avoid accountabilities and liabilities leads to excessive efforts (i.e., striving for some sort of perfection) in monitoring and related data analysis. Monitoring gives insight into the situation at hand. But getting to know all the details that could potentially be relevant might be in conflict with focussing on the essentials. When people are obliged to constantly demonstrate that they are doing their work rightly, it will induce fear of making mistakes and affect their capacity to execute the work in the first place. Getting the job done is where most organizations earn their money and generate their societal impact. The most well-known examples are found in education and health care, where teachers and health professionals sometimes spend more than 20% of their time on administrative activities which are supposed to contribute to better quality and safety. The effects are, however, that there is less time for education or health care and that work pressure is increased which is certainly *not* contributing to better quality. Moreover, teachers and health care workers do not choose their profession to do administrative tasks; these affect the motivation for their work as well as their wellbeing

Similar unbalances between perfection and good enough, though perhaps less pronounced and less visible, are to be found in many commercial industries. In SHW management, there can be the well-known unbalance between 'paper safety' (i.e., being able to show on paper that things are perfectly all right – often referring to compliance) and real safety. All too often, SHW management systems are associated with a significant amount of bureaucracy and an excess of procedures and prescriptions. This is the dominance of what Hale and Borys called 'the top-down classical approach to rule making', while it is important to reconcile that view with a bottom-up constructivist view on rules.[3] It is good to realize that rules are not by definition prescriptive. There are rules in the form of performance goals (such as ALARA), and there are process rules that define the process that should be followed (leaving considerable freedom for how that is done).[4] An interesting concept is that of 'red rules', i.e., rules that always should be complied to (e.g., because of the immediate danger of serious irreversible impact), while normal rules are not that critical.[5]

The balance between efforts to get good risk assessments and actions to improve SHW is another example. The effort of carrying out risk assessments that address every SHW risk, even minor ones, may easily deplete the energy to deal adequately with the most important risks. Governmental inspectorates and certification bodies may, unintendedly, foster such unbalances, thereby frustrating the process of continual improvement. For organizations and institutions, the right goal is realizing significant improvements in practice, not being complete or realizing improvements on paper.

8.1.2 Innovation

Innovation is the other broadly accepted strategy for achieving significant long-term improvements. A case study about the New Zealand Aluminium melter[6] describes how over a range of years various innovations in production technologies improved safety to the degree that it was known as the world's safest aluminium melter. From a system perspective, it is useful to distinguish innovation within the existing (socio-technical) system and innovation of the system. Vision Zero implies a range of innovative perspectives (see Chapter 3).[7]

8.1.2.1 Innovation in the System

The technology can be replaced by a similar but more efficient technology without changing the system; for example, when a motor is replaced by a more energy-efficient one. The organization can be improved without drastic changes, for instance, by broadening the qualifications of people (e.g., mechanical engineers are also qualified for electrical tasks). In the example that gives greater flexibility in the planning of personnel. These are examples of innovation in the (existing socio-technical) system. Innovation in the system implies minor but significant innovations whereby the socio-technical system improves while it remains largely unchanged. Innovation in the system actually contributes to optimization.

8.1.2.2 Innovation of the System

The business analyst Singh introduced the term 'Innovating to Zero' (see Section 3.6). According to Singh,[8] Innovating to Zero is not a problem-solving strategy (like optimization), but an alternative way of running and innovating one's business. It can be applied not only to SHW but also to many other 'zero ambitions', for instance, defects. Of course, such ambitions sketch a world that is almost utopia, and achieving zero often sounds hardly possible. But Singh states that the journey is much more important than the destination, and moving towards zero goals implies creating huge challenges and opportunities. *When zero is the ultimate goal, then even achieving its objective partially will be huge progress and make a real difference to society.*[9] *He also states that 'Success in innovating to zero requires an innovation agenda that bravely talks of breakthroughs that intend to create a better world, a zero concept world, which is free of unhelpful externalities and defects. It needs a strong culture from people within that ecosystem'.*[10]

The theory U[11] describes the innovation process as a transition process in U form. It is basically a process of gradually understanding deeper assumptions that underlie our thinking of current reality. By reflecting on the purpose, the turning point can be reached. Then after reframing our principles and developing new thinking, new activities and processes can be created leading to new structures and practices. In the U-process, a certain step back is needed before a jump forward can be made. Consequently, Vision Zero is not always a matter of linear progress, sometimes it may take time and preparation before the next significant improvement can be realized.

A very relevant concept is that of *workplace innovation.*[12] Workplace innovation can be regarded as the complement of technological innovation. It comprises innovations in the work organization, in human resource policies and labour relations in the workplace. Participation as well as better utilization of human talents (of both workers and managers) are thereby core elements. The workplace is a dynamic environment wherein technological and social innovations are important for significant improvements.

8.2 UNDERLYING AND SUPPORTING ISSUES AND PROCESSES

In this section the focus is first on the following supporting processes: Legal requirements for SHW risk management, standardization, adapting to the state of the art, Deming's Plan, Do, Check, Act cycle, reducing human error, developing a prevention culture and learning.

8.2.1 Legal Requirements for SHW Risk Management

The idea of improving SHW is a common feature of legal SHW frameworks in many jurisdictions. Most jurisdictions thereby use a combination of a prescriptive approach with a goal-oriented or risk-based approach. For instance, in the European Union's Framework Directive, employers are legally obliged to assess the SHW risks and to take adequate measures to eliminate or reduce SHW risks. This implies a step-by-step approach to improving SHW; the underlying assumption is that the SHW performance will be improved over time till it is fully optimized. It takes time to get rid of all existing problems in a production system, often several years. In a stable situation, every SHW problem that is solved will automatically imply better SHW performance and bring you one step closer to the ideal optimum situation. For many decades, production systems and their organizational environment were basically stable. It therefore made perfect sense that the legislator asked organizations (implicitly) to optimize SHW in their production system. However, since the 1980s innovations and change have accelerated enormously, which greatly affects processes of stepwise improvement. In an era of dynamic change, stepwise improvement implies a continuous struggle with a changing working environment: new challenges come forth regularly, and it becomes a challenge to avoid *running behind the facts*. The situation is dynamic; while problem A is solved, problem B pops up. Steps forward are sometimes followed by a step back. The process of incremental improvement is then easily frustrated. In changing circumstances, the optimum is also changing. Striving for the optimum becomes something like trying to hit a moving target. This makes it difficult to realize the optimum; and in case it is realized, the optimum will be outdated at short notice.

The legal requirements for risk management are based on the risk control strategy. It implies that SHW risks get priority depending on their severity; this has some inherent limitations. After the most pressing SHW issues are solved, the remaining SHW issues are likely to get lower priority, while minor issues will get less or no priority. It is easy to understand that progress towards better SHW will then retard or eventually stop at all. It was already mentioned that for Vision Zero, with a view on long-term continual improvement, a commitment strategy is to be preferred over the risk control strategy. See Section 3.7

8.2.2 Standardization

Standardization was first introduced in the technical sciences. Standardization for safety has a long history. The International Electrotechnical Commission (IEC) responsible for standardization of electrotechnical products installed its first technical commission devoted to human-machine safety in 1926.[13] Standardization helps to improve the predictability of processes and to avoid unnecessary deviations. Natural scientific laws determine technical phenomena, and it is possible to determine objectively and reliably whether a technical issue meets the requirements of the standard or not. Standardization of inputs, production processes, ways of working and tools implies the aim to reduce variation. In this way the likelihood of problematic deviations and of accidents is greatly reduced; only minor acceptable risks should remain. But there remains always the (relatively unlikely) possibility of phenomena that were regarded as irrelevant in a certain context can be relevant in another context.

That is why it is important that standards are evaluated periodically, for instance, every five years, and adapted when necessary.

Today standardization is also applied to organizational issues and work processes. Any rule, protocol or prescription is a form of standardization. Again, standardization can make organizational processes and work processes more predictable and controllable – up to a certain level.

Researchers use their models to get a grip on technologies, organizational factors and to some extent people. Models reflect dynamic situations and are increasingly used to create more predictable situations: models often become de facto standards. In complex situations 'digital twins', i.e., digital models that are used to explore and predict the behaviour of complex (socio-technical) systems, are used.

Formal standards are defined by stakeholders and experts. The involvement of experts means that their expertise becomes available via the standards to a much larger audience, which is very positive. It is important that standardization committees represent all relevant interests in a balanced way. This is often not fully the case. The required investments in time and money for participating in standardization committees form a hurdle for participation, especially for less financially powerful stakeholders. Standards sometimes reflect primarily the business interest of powerful players on the market, rather than the shared interest of all stakeholders. Some areas that are standardized are very specialized. Different standards for closely related topics may then include contradictory requirements. This may lead to confusion or a loss of oversight for the users.

Even an organization that has perfectly standardized its technology and its organizational processes still has to deal with a changing organizational environment and with people. The organizational environment cannot be kept fully under control or standardized. External developments or unexpected events can disturb standardized processes. For work carried out in the open air (e.g., construction), the weather conditions (storm, rain, snow, very hot or cold temperatures, etc.) may affect work and production processes. Time may wear out technological assets (ageing of technological assets) or affect people (ageing of people). Human behaviour, learning and creativity cannot be standardized, while interpersonal interactions (communication, cooperation, conflicts, etc.) are by their very nature ambiguous and can at best be partly standardized.

While people and their behaviour cannot be standardized, a certain degree of standardization is possible. The required competencies can for instance be standardized by defining adequate qualifications. Work can be standardized to a certain extent: working times (e.g., nine to five, or in shifts) and remuneration of work can be standardized per job type, etc. The ways of working can be prescribed in procedures and prescriptions. The more workers are regarded as *human resources* (not as individual people) the more standardization can be expected (all too often still based on the white male standard). Sophisticated tools for human resource management make use of all kinds of big data and *people analytics* to further standardize and control human requirements and individual performance. It is thereby neglected that individuals are unique and that team performance is likely to be more important than individual performance.

A good cook does not need a cook book. Though cooks may use cook books as inspiration for new ideas, cook books do not help to improve their cooking. But for someone who is inexperienced in cooking, a good cook book is almost indispensable. More generally the question is: what is the added value of standardization? The answer is that compliance with standards can be objectively verified and that is essential for smooth co-operation between different organizations, people and cultures. Complex products can only be made reliable due to a high degree of standardization of materials, tools and processes. Standardization may contribute substantially to improving safety by reducing the likelihood of deviations that may lead to incidents and accidents. But professionals – like good cooks – may not like to work according to standards and experience them as mistrust of their professionality.

Making a new protocol, rule, prescription or on the societal level new legislation is often a first impulse after an incident, the so-called *risk-regulation reflex.*[14] It is often regarded as the solution for a new problem/challenge an organization is confronted with. However, a quickly made new rule is likely to focus on the symptoms rather than underlying causes and is likely to have limited added value. Many organizations and professionals suffer from too many (external as well as internal) standards in the form of rules, protocols or prescriptions. The added value of each individual standard is not always clear. People can only understand and memorize them up to a certain level. Moreover, each standard must be regularly updated, and compliance to standards may easily trigger a culture of compliance and ticking the box. It is therefore important to reflect on the need for new rules or standards before developing or adopting them. It is a general phenomenon that is much easier to develop a new standard or rule than to terminate them. Indeed, it is much easier to make things more complex, than to make them simpler.

8.2.3 Adapting to the State of the Art

The state of the art in technology and organization develops over time, due to scientific progress and innovation. That offers opportunities to overcome the limitations posed by existing technologies or ways of organizing. This implies opportunities for improvement that go beyond merely problem-solving. The improvement of the production system and of SHW can get a boost by adopting more efficient or safer and healthier technologies or ways of working, without requiring drastic changes in the socio-technical system as such. Both industrial guidelines and standards may refer to the best available practices. The term *best practice* can be useful when referring to technologies, but for socio-technical systems performance depends partly on the context, and different technologies may be 'best' for different contexts. The term *good practice* seems then to be more adequate.

Requirements for adapting to the state of the art can be included in legislation. That has led to terms like *Best Available Technologies* (BAT) or *Best Available Technologies Not Exceeding Excessive Cost* (BATNEEC) and *As Low as Reasonably Practicable* (ALARP). The intention is to trigger organizations to innovate and to phase out older, less desirable technologies. But in reality, it is difficult to identify the Best Available Technologies in an objective way, while today's best technologies may no longer be the best technology tomorrow. Best technologies often serve as a kind of informal or alternative standard. Adapting to the state of the art can be conducive to optimization as well as innovation.

8.2.4 The Deming Cycle: Plan, Do, Check, Act

Management system standards of the ISO type suggest that implementation of those standards will generate a process of continual improvement. The ISO 45000 series focuses strongly on the control and assurance of organizational processes. These processes should be well-defined, documented, implemented, audited and periodically reviewed. The iterative Plan, Do, Check, Act (PDCA) process is meant to support the continual improvement by 'raising the bar' of SHW performance a bit higher in every PDCA cycle, usually annually. It is certainly useful – at least for larger organizations – to implement an ISO 45001 type management system, as it helps organizations to work more systematically in managing occupational safety and health. Perhaps surprisingly, scientific research has not led to convincing evidence that these management systems lead to continual improvements in SHW performance, nor that organizations that have certified SHW management systems outperform similar organizations (same sector, size and country) without such certificates. There are six main reasons that explain these unexpected and unintended results.

1. The focus on control and assurance reflects the *risk control strategy*. The risks are assessed, and when risks are regarded unacceptable the necessary measures for risk reduction and control will be taken. As a result, the highest risks will be reduced, and minor risks will remain. In the PDCA cycles thereafter, the remaining risk will be minor. As minor risks usually get a lower priority, every cycle means ever lower priority for solving the remaining SHW problems, and as a consequence fewer, if any, improvements will be realized.
2. Documented procedures are an essential element in ISO-type management systems. Such procedures are almost by definition somewhat generic as they have to be relevant for a range of similar situations in the organization. They cannot be tailor-made for each and every specific situation. As a consequence, there is often an unintended difference between the actual way in which the work is carried out (Work as Done – WaD) and the way the work is prescribed (Work as Imagined – WaI). There are limitations to the improvements that can be achieved by the implementation of documented procedures or prescriptions. Procedures can allow the workers more control over their work than prescriptions.
3. The phases of Plan and Check have inherent limitations. As the future is unpredictable, planning has its limitations, especially in dynamic or complex situations. For the phase Check, it is important to measure or monitor performance. However, tangible factors are easier to measure than intangible factors. Not all factors can be quantified (e.g., trust). What gets measured gets done, but the other side of the coin is that what is not measured can easily be

neglected. It helps when the outcomes of measurements are used as input for meaningful conversations on what (further) improvements could be possible.

4. In an ideal stable world, each PDCA cycle implies opportunities for reaching the next step in SHW, as in the process of optimization. Again, we live and work in a turbulent world of ever-increasing change. There are changes in the organizational context (economic, political, technological, ecological and cultural), as well as internal changes (e.g., changes in leadership, strategy, profitability, structure, personnel). All these factors will to a certain degree influence SHW processes and performance. That is why the intended process of continual improvement often becomes unintendedly a process of continually *keeping up with change*.
5. For ISO 45001 certified management systems (and similar ISO management systems for other areas), to have the certificate can easily become an aim in itself. For senior leaders, it can be a comfortable idea that there is periodically an external check on the functioning of the management system. External pressure to demonstrate compliance to an ISO standard may play a role. When having the certificate becomes the aim, continual improvement is reduced to compliance with the standard. This then becomes more important than the organization's desire for further improvements.
6. External auditors of SHW management systems are often experienced ISO 9001 auditors who qualified themselves for SHW as well, but without having working experience as an SHW expert or practitioner. They are likely to have a limited understanding of Work as Done and focus mostly on Work as Imagined. SHW risks can differ greatly among different sectors. Auditors who have experience with SHW are competent in sectors they know well, but in other sectors they might have problems to identify SHW risks or their underlying factors.

Together the reasons mentioned above explain why the aim of continual improvement over longer periods of time is difficult to realize with ISO-type SHW management systems, whether certified for ISO 45000 or not. Only in the first years of implementation, significant improvements can be expected as a result of working more systematically and having clear priorities for reducing the most important risks. Thereafter performances are likely to level off; learning and the development of a prevention culture are the key factors for further improvement of SHW performance.

There are two ways in which the PDCA cycle can get an increased potential for improvement. Deming, the name giver to the PDCA improvement cycle, emphasized the importance of studying the results after 'do'. He judged it essential to analyse and understand the results obtained before taking action. In his later works, he therefore referred to the Plan, Do, Study, Action (PDSA) improvement cycle. By emphasizing the study aspect, the Deming cycle becomes very similar to Kolb's learning by doing the process of Observing, Analysing, Decision-making and Doing[15] (see Section 4.7). Another way to increase the potential of ISO-type management systems is to introduce the proactive leading indicators (see Section 9.2) to steer the management systems in a more proactive way. When the management systems are no longer based solely on the risk reduction strategy, documented procedures can be reduced due to increased ownership and competence. Keeping up with changes is minimized because of earlier anticipation, while certification, i.e., compliance to international standards becomes less important compared to internal standards and expectations. It implies that internal audits and reviews become more important compared to external audits.

8.2.5 Human Error and SHW Behaviour

As technology and organizational processes are easier to control than human behaviour, people are often seen as the weakest link in the management of SHW. It is therefore not a surprise that it is popular to focus on human behaviour and to try to reduce human errors. That makes some sense, but often it has only limited value in practice. Though short-term success may be achieved, focussing on human behaviour and trying to reduce human error are not successful strategies for continual long-term improvement. It is usually focused on compliance with procedures and prescriptions, possibly while using sanctions. This means those strategies are strengthening external motivation, and unfortunately that will destroy intrinsic motivation. Consequently, it will not help to realize the aspects of Vision Zero that are based on intrinsic motivation and ownership (leadership, proactivity, creativity, ownership, higher-order learning, etc.)

In attempts to reduce human error, there is often a strong emphasis on reducing human error at the operational level. The underlying idea is that it is the sharp edge where most critical errors are made. That assumption is wrong. Most human errors are triggered by the way the work is organized, by the way tools and machines are designed, by having insufficient information about specific hazards and risks, by time pressure, etc. This implies that the decisions of managers and designers are the main determinants of human error. Improving leadership, work organization and the design of technology and workplaces often make much more sense than focusing on worker behaviour. Kletz said:[16] 'try to change situations, not people'.

It is human to make errors. Instead of focusing on the prevention of human errors, it makes more sense to make sure that errors cannot have *serious impacts* on SHW. This requires work processes with checks and balances, for instance, the application of the four eyes principle in the case of SHW critical activities and a culture wherein errors are tolerated but learning from them is seen as essential. Buffers and redundancies may help to minimize the impact of human errors.

In 'the new view on safety', human error is sometimes regarded as a potential source of improvement and innovation. That is wishful thinking. The impact of human error is rather random and usually it does not contribute to planned action; only by exception it may lead to improvements (serendipity). A more useful strategy is to increase the capacities of people to deal with SHW challenges and problems: SHW education and training, as well as dedicated safety and health promotion. Making people more SHW-competent does not only directly improve their knowledge but can influence their behaviour. It is conducive to more fruitful dialogues between colleagues, also cross-sectional and across-hierarchy. That will stimulate mutual learning and SHW improvement.

Another way to address the behaviour of people is via the development of a prevention culture. The idea is that by setting standards for desirable behaviour and discouraging undesirable behaviour, the people's behaviour will improve. It is assumed that by subtle forms of social control, the behaviour will become more predictable. The focus on promoting desirable behaviour and discouraging undesirable behaviour fits very well in an optimization strategy. Learning is then mostly limited to first-order learning.

8.2.6 Creating Synergies

In Chapter 3 on Vision Zero, the importance of synergies was already highlighted. By creating synergies between different aspects or processes, the resulting outcome is more than merely the sum of the two individual factors. In this way, performance can be improved. Synergies can be created between factors within the system (e.g., SHW), contributing to the optimization of the system. Synergies can also be created between factors that were not considered as part of the same system (e.g., synergy between SHW management and business processes). By involving factors that were initially seen as outside the system (e.g., beyond the SHW management system), synergies imply an enlargement of the system; this may contribute to optimization as well as innovation (see Section 10.1).

8.2.7 Economic Incentives

Financial aspects play a role in almost any decision. It is then a logical idea to use economic incentives to influence the behaviour of people as well as organizations. However, that often turns out to be a double-edged sword. When the aim is to influence the behaviour of organizations, it is important to realize that many costs of poor SHW do not weigh on the organization. Several costs are usually shifted to society at large, e.g., cost of health care, disability and early pensioning; other costs are shifted to the individual workers and their families. Usually, there are perverse incentives that discourage investments in prevention. It is in itself a sound idea that organizations should bear the true costs of poor SHW. However, it remains a problem that the costs of SHW are much more tangible than the SHW benefits. Introducing economic incentives may strengthen in organizations as well as individual people the calculative attitude of *what's in it for me*. But SHW is not a product on the market that should be bought for the best price. This may withhold organizations to acknowledge the value of SHW. For individual behaviour, it is known that financial rewards, as a form of external motivation, may work to some extent – especially on the short run. Unfortunately, money is a satisfier that easily loses its motivating impact. Moreover, economic incentives and rewards seriously undermine intrinsic motivation.

8.3 OPTIMIZATION AND INNOVATION (II)

Innovation in the system contributes to optimization, innovation of the system implies more far-ranging changes. Table 8.1 gives an overview of the characteristics of the two key concepts for long-term continual improvement.

TABLE 8.1
A comparison of optimization and innovation

Item	Optimization (including innovation in the system)	Innovating of the System – Innovating to Zero
System level	Improvements in the socio-technical system	Improvements of the socio-technical system
	No change in organizational identity	Organizational identity may change
Leadership	Transactional and transformational leadership	Visionary and transformational leadership
Main goals	Operational excellence	Benefits of novelty
	Compliance	Beyond compliance
	Efficiency	Efficacy
Time scale	Continual short-term improvements	Significant impacts on the longer term
Relevance	Relevant for each organization	Mostly found in young and smaller organizations
Interactions	Optimization does not require innovation but can be strengthened by innovation in the system	Innovation requires optimization as complementary strategy (as part of a maturing process)
Characteristic processes	Continuity and gradual development	Breakthrough and disruptive change
	Control and a predictable impact	Experimentation, learning and less predictable impact
	First-order (adaptive) learning	Higher-order (innovative) learning; learning to learn
	Focus on best practices in the organization and frontrunners in their own sector	Focus on best practices from frontrunner organizations, including those from other sectors
Synergies	Synergies between factors in the system	Synergies between internal and external factors
Impact	Impact mostly internal	Internal as well as external impacts (including creative destruction)
	Involving traditional stakeholders	Building new stakeholder relationships
	Introduction of new risks unlikely	Possible introduction of new hazards and risks
The role of SHW experts	Strong focus on problem-solving and optimization	Focusing on promoting SHW innovation to zero as a complement to problem-solving and optimization

8.4 CONCLUSIONS

Optimization and *innovation* are the two main long-term strategies; both are important for Vision Zero. Optimization processes are characterized by control, efficiency and adaptive learning, while innovation is rather a matter of experimentation, novelty and innovative learning.

Optimization, like problem solving, is relevant for any organization. In theory, innovation is relevant for each organization too, but it is most important for young and smaller organizations as they have fewer vested interests in the existing production systems and markets; they often have to be innovative in order to be able to compete with larger organizations who have the benefits of economies of scale.

Optimization does not affect the main characteristics of the existing production system. In the short term, optimization is a strategy with only winners; investments are relatively minor, and improvements seem guaranteed. Optimization is unlikely to introduce new significant hazards and risks, while innovation implies uncertainties and may introduce new hazards or risks.

But innovation may lead to systems that are significantly better both for SHW and productivity. Innovation does require more investments and is associated with greater business risks. Innovations often imply that there are 'teething problems' at the beginning, which means that the new production system is far from being optimized. Indeed, after radical innovations, the process of optimization will help to improve the system further. Innovation and optimization are complementary in this respect.

In the longer run, innovation to zero may lead to significantly better performances in SHW and productivity. Optimization aims at operational excellence, compliance and efficiency. Innovation aims primarily on the benefits of novelty, and the focus is thereby on efficacy rather than efficiency. The long-term process of innovation to zero is not a smooth process of linear continuous development, but a process with disruptions that imply that existing arrangements become obsolete. There is then a certain discontinuity, and there are vested interests that will resist these types of radical innovation.

Several important concepts and theories that underlie Vision Zero support optimization as well as innovation processes. Optimization and innovation both require leadership, perseverance and investments in prevention (see Chapters 5 and6). Proactivity, commitment, motivation, ownership, shared values, learning and reflection on mental models may support either of the two strategies but are most important for innovation.

NOTES

1 Weick, K.E., Sutcliffe, K., (2007) *Managing the Unexpected.* San Francisco: Jossey-Bass.
2 Compare with Imai, M., (1986) *Kai Zen.* New York: McGraw-Hill.
3 Hale, A., Borys, D., (2013) Working to rule, or working safely? Part 1: A state of the art review. *Safety Science* 55, 207–221. Doi: 10.1016/j.ssci.2012.05.011.
4 Hale, A., Borys, D., (2013) Working to rule, or working safely? Part 1: A state of the art review. *Safety Science* 55, 207–221. Doi: 10.1016/j.ssci.2012.05.011.
5 Yates, G.R., Hochman, R.F., Sayles, S.M., Stockmeier, C.A., (2004) Sentara Norfolk General Hospital: Accelerating improvement by focusing on building a culture of safety. *The Joint Commission Journal on Quality and Safety,* 30 (10), 534–542.

6 Young, S., (2014) From Zero to Hero. A case study of industrial injury reduction: New Zealand Aluminium Smelters Limited, *Safety Science*, 64, 99–108. Doi: 10.1016/j.ssci.2013.11.016.
7 Zwetsloot, G.I.J.M., Kines, P., Wybo, J.L., Ruotsala, R., Drupsteen, L., Bezemer, R.A., (2017) Zero Accident Vision based strategies in organisations: Innovative perspectives, *Safety Science*, 91, 260–268. Doi: 10.1016/j.ssci.2016.08.016.
8 Singh, S., (2012) *New Mega Trends – Implications for Our Future Lives.* Hampshire, UK; New York, US: Palgrave Macmillan.
9 Singh, S., (2012) *New Mega Trends – Implications for Our Future Lives.* Hampshire, UK; New York, US: Palgrave Macmillan.
10 Singh, S., (2012) *New Mega Trends – Implications for Our Future Lives.* Hampshire, UK; New York, US: Palgrave Macmillan.
11 Scharmer, C.O., (2007) *Theory U – Leading from the Future as It Emerges.* Cambridge, MA: Society for organizational Learning (SoL).
12 Eeckelaert, L., Dhondt, S., Oeij, P., Pot, F.D., Nicolescu, G.I., Webster, J., Elsler, D., (2012) Review of workplace innovation and its relation with occupational safety and health, EU-OSHA, Bilbao. And Oeij, P.R.A., Ruis, D., Pot, F.D., (2017). *Workplace Innovation – Theory, Research and Practice.* Cham, Switserland: Springer.
13 IEC, (2020) *White Paper: Safety in the Future.* Geneva: International Electrotechnical Commission. Available at https://webstore.iec.ch/en/publication/67876 (Accessed 17 December 2024).
14 People easier add new thinks than making things simpler. See: Adams, G.S., Converse, B.A., Hales, A.H., Klotz, L.E., (2021) People systematically overlook subtractive changes. *Nature*, 592, 258–261. Doi: 10.1038/s41586-021-03380-y.
15 Kolb, D.A., (1984) *Experiential Learning – Experience as a Source of Learning and Development.* New Jersey: Prentice Hall.
16 Kletz, T., (2001) *An Engineer's View on Human Error* (3rd edition, first edition 1985). Rugby, UK: Institution of Chemical engineers.

9 Strategies for Practical Implementation of Vision Zero

Vision Zero is only a useful concept when it is implemented and triggers processes of commitment and improvement. In earlier chapters, several examples of approaches or tools were already presented because theories and practices should go hand in hand. In this chapter practical implementation strategies are the focus. The first two sections, on the seven Golden Rules and the 14 Proactive Leading Indicators, were specifically developed for implementing Vision Zero, though they can be used by organizations that have not adopted Vision Zero too. The other approaches are rather innovative methods for SHW improvement, also in a long-term perspective, important for Vision Zero. It is good to keep in mind that in a socio-technical system, there are numerous relevant relationships between what seem to be independent elements. For reasons of readability, only some of these relationships are explicitly mentioned in this chapter.

9.1 THE SEVEN GOLDEN RULES

Any organization can adopt Vision Zero to improve SHW. It is not only for the frontrunners in SHW but also for companies who want to get rid of all their SHW incidents. Many companies that adopt Vision Zero can improve their performance significantly by implementing some state-of-the-art practices including more traditional approaches that focus on compliance. But it is an essence of Vision Zero that the ambition of the organization is to go beyond compliance. How then to combine these apparently contrasting approaches in a useful and inspiring approach for Vision Zero?

It is exactly for this paradoxical challenge that the ISSA developed, in an interactive process with more than 1000 companies from all continents, led by Helmut Ehnes, the seven Golden Rules.[1] Each of the seven Golden Rules builds on good practices and is focused clearly on SHW prevention and the promotion of SHW improvement processes. The Golden Rules are compatible with OSH legislation but also imply some next steps. They put the focus on seven different aspects that jointly form a systemic approach to SHW improvement, combining traditional and innovative perspectives. The seven Golden Rules differ from traditional rules because they do not prescribe activities but address SHW from seven complementary points of view. It is up to the organization and its leaders to use these seven perspectives in a fruitful way for self-assessment, reflection and implementation; the ISSA guide

DOI: 10.1201/9781003621225-9

The seven Golden Rules of Vision Zero for Safety, Health and Wellbeing at work	
1	Take leadership – demonstrate commitment
2	Identify hazards – control risks
3	Define targets – develop programmes
4	Ensure a safe and healthy system – be well-organized
5	Ensure safety and health in machines, equipment and workplaces
6	Improve qualifications – develop competence
7	Invest in people – motivate by participation

FIGURE 9.1 The ISSA seven Golden Rules. (The 7 Golden Rules stem from ISSA, (2017). The Seven Golden Rules. Geneva: International Social Security Association. https://visionzero.global/all-documents?title=&field_category_target_id=458&field_publication_date_value= (Accessed 19 December 2024).)

provides useful guidance.[2] The seven Golden Rules are developed for the implementation of Vision Zero, but they can also be used by organizations that have not (yet) adopted Vision Zero. An overview of the seven Golden Rules is given in Figure 9.1.

The seven Golden Rules (GRs) are not independent; they interact and strengthen each other. For instance, leadership (GR1) is important for all other Golden Rules.

The innovative character of Vision Zero is mostly represented by GR 1(Take leadership – demonstrate commitment) and GR 7 (Invest in people – motivate by participation). These areas are usually not, or only implicitly addressed in OSH legislation or traditional OSH management approaches. The impact of GR 3 (Define targets – develop programmes) can also be innovative, but it is up to the organization to define programme and targets, which can range from very innovative to rather traditional.

The four other Golden Rules build on proven good practices and are fully in line with broadly adopted legal requirements. The idea is not that the seven Golden Rules should be prescribed, but rather that companies adopt them as important areas or principles for strengthening prevention and developing a prevention culture. It is up to senior leaders to adopt the seven Golden Rules. The commitment of senior management is essential for guiding the preventive efforts of the organization and for the long-term perseverance required to make Vison Zero a success. The ISSA guide on the seven Golden Rules is therefore primarily meant for senior managers and SHW professionals who operate in a strategic way jointly with senior management. The guide helps top managers to self-assess the organization's preventive efforts and gaps therein. The seven areas addressed are, however, also very useful for other target groups that are interested in promoting excellence in SHW, e.g., policy makers, inspectors and worker representatives. Below the seven Golden Rules are shortly presented and discussed; for more details, it is best to consult the ISSA guides.[3]

9.1.1 Golden Rule 1: Take Leadership – Demonstrate Commitment

Senior managers form the main target group for the seven Golden Rules. The importance of leadership for Vision Zero was already discussed in Chapter 5. In the ISSA guide on the seven Golden Rules, GR 1 focuses on four aspects of leadership: (1) to give direction to preventive efforts, by defining goals and ambitions and setting standards. (2) To communicate and make sure that everybody in the organization knows that SHW is important, by promoting transparency and creating a culture wherein everybody is informed, and it is always possible to discuss SHW issues. (3) To emphasize the importance of SHW in all activities and for all people. (4) To make sure that supervisors as well as workers have the necessary resources (time, money, tools, knowledge) to contribute to SHW. Several additional relevant aspects of leadership are described in Chapter 5.

9.1.2 Golden Rule 2: Identify Hazards – Control Risks

Identifying hazards and risks and evaluating the control thereof is a key issue for OSH management. It is important to start this already in the early stage of design.[4] The technology chosen, the design of the work organization and of the work processes and work place determine the hazards and risks at work to a high degree. In practice however, risk assessments are usually carried out in the production stage. Then, changes in the technology, work organization, work processes or workplaces are relatively expensive. Consequently, many organizations and SHW professionals do not consider such changes as feasible and focus on less effective added-on preventive control measures (see Chapter 6).

It is important to involve people who know the reality of the work processes (operational workers) in doing a risk assessment, even in the design stage. It is important that risk assessments are documented but do not make it a paper exercise. Information about the risks should be shared with the workers and supervisors of the relevant workplaces; when relevant they should also be shared with personnel of contractors, temporary agencies, etc. A risk assessment is no aim in itself but makes it possible to anticipate accidents, illnesses and the associated production losses. They should always be followed up with improvement actions.

Risk assessment should be complemented by learning from events or situations that turn up in practice, such as incidents, near misses, illnesses and SHW complaints. These events indicate that not all risks are properly controlled and shed a new light on the risks in the workplace; Such events require follow up and form important feedback that should be used to improve or update the risk assessment. As set out in Chapter 6, it is important to give feedback also to those involved in the earlier stages of development and design. In these ways, risk assessments can contribute to the development of a learning prevention culture.

9.1.3 Golden Rule 3: Define Targets and Develop Programmes

Vision Zero is the expression of the organization's ambitions in SHW. It is a logical consequence to have dedicated (temporary) programmes with clear targets to improve specific SHW areas. The focus can be anything and may range from better SHW procurement or better maintenance to the reduction of physical load or noise, or the reduction of stress. Programmes can also be directed at underlying factors like social skills, leadership or a culture wherein everybody feels free to discuss sensitive SHW issues. Programmes are especially helpful to improve in areas that hitherto did not get sufficient attention, or to new and emerging issues that were not yet top of mind. Given the crucial role of leadership in the success of Vision Zero, consider always the option of an SHW leadership development programme.

Any SHW programme should have clear goals or targets. That implies the opportunity to monitor progress and evaluate effectiveness. In evaluating programmes attention should not only be given to the achievements realized but also to the relevant processes, i.e., what was successful, or not, what triggered enthusiasm, what triggered resistance, what were unintended side effects, etc.? What can be learned from the implementation of the programme? Keep in mind that for the Vision Zero ambition, the long-term impact of the programme is key. It is often useful to apply the SMART principle, but sometimes that can drive away attention from more essential issues, for instance when the programme is focussed on intangibles like sharing values or promoting a culture of care. Check with the proactive leading indicators (next section) of any of these can be helpful to set targets for the programme.

The commitment strategy for Vision Zero makes it preferable to involve people from the shop floor in SHW programmes. The *be early* principle implies that some form of participation is often already useful in the phases of goal setting and design. It is also useful to involve workers in the evaluation of programmes.

9.1.4 Golden Rule 4: Ensure a Safe and Healthy System – Be Well Organized

GR 4 is about ensuring a safe and healthy system and being well organized. A system approach is important here in several ways. First, it is important to take prevention measures at the *system level*, instead of focussing on specific incidents or human errors.[5] Only changes in the socio-technical system influence *patterns* of events.

SHW prevention should be organized systematically as well. For large and medium-sized organizations this can be done via implementing SHW management systems. Different from the production system itself, a management system is a so-called aspect system: it is solely the SHW aspect that is managed systematically. Management systems are often useful for compliance with legal obligations and customer requirements. A management system can help to define accountabilities, but also to delegate responsibilities in a right and just way to lower hierarchical levels. Dedicated attention to the design stage should also be included. A management system is, however, more than just working systematically: it is regarded as a whole (a system). A management system is periodically audited to ensure it is still functioning as intended. Periodic management reviews are meant to adapt the system when needed so that it still makes sense in changed situations. Make sure the associated paper work is kept to a minimum. SHW performance is realized in daily practice, never on paper. Certification of a management system may help to increase management support for SHW, but the certificate may become a goal in itself; that easily undermines further progress in SHW performance.

9.1.5 Golden Rule 5: Ensure Safety and Health in Machines, Equipment, Materials and Workplaces

The focus of GR 5 is on the technological aspects of the production system. The use of technical items and materials is often associated with hazards and risks. Prevention is relevant both to hardware such as machines, equipment, materials and workplaces, as well as to software. Start with attention to SHW in the selection, development, design and procurement of production technologies and tools. Use thereby the 12 principles for Worker-Friendly Design, see Section 6.5.5. In the phase of production and maintenance, technical and organizational measures to control SHW risks are important. This usually requires well-planned preventive maintenance and adequate competencies of those who work with the technologies. Regular troubleshooting is a clear sign that preventive maintenance should get higher priority.

Make sure to comply with important technical standards (international such as those of ISO or IEC, or standards of the organization). Keep an open eye on technological developments in other sectors and regions to ensure the best available technologies are chosen.

Technological hazards are not only relevant for safety but just as well for health and wellbeing (substances, dusts, noise, biological materials, etc.). Occupational exposures are often difficult to evaluate and may require the involvement of dedicated experts, e.g., of occupational or industrial hygienists or radiation experts. Ensure that devices for safety and health are always available and function properly.

As technological items are designed by engineers with a technological focus, technological items are not always user-friendly. It is important to involve ergonomic expertise to look at the human-technology interfaces in workplaces and of tools. This will help to ensure that tools are easy to use in the right way, and that the physical and mental burden of work are reduced.

9.1.6 Golden Rule 6: Improve Qualifications – Develop Competence

People play an important role in any organization. Their commitment is key to improving SHW. But people can only work adequately when they are properly educated and qualified for their jobs. Managers and workers should be well-qualified and trained in SHW matters. Only then they are able to know and understand the SHW opportunities and risks, and how to deal adequately with them and prevent SHW problems. It is important to repeat training at regular intervals. Take care that the development of competences follows the rapid developments in the state of the art of technology, rising societal expectations and legal requirements. Support the development and personal growth of the personnel – that is also important for innovation and for coping with change. Learning on the job is an important complement to training and education. Make sure the work organization is designed in a way that learning by doing will result naturally. This requires sufficient possibilities for the workers to control their work processes. It is also essential that workers receive feedback and social support from co-workers and their supervisors, see Section 4.7.

Training and education should not be too abstract. Learnings should be easily applicable in work practice. Blended learning, wherein formal training and learning on the job alternate can be a good idea. The linguistic skills of the people should be taken into account. Some may have difficulties in understanding the training; immigrants may have limited skills in what is not their native language. Others may have difficulties with reading, especially texts that use jargon or difficult words.

Usually, most attention goes to the education and training of the workers at the sharp end. But it is at least as important to qualify and periodically train the (top) managers: they have a significant influence on the SHW in the organization. Training and education should be interactive and focused on the target group. Acknowledge the tacit knowledge of the workers and use it to make the training better applicable in practice. Listening to the practical wisdom of operational workers is often more valuable than instructing. Remember that *Work as Done* is always more important than *Work as Imagined.* Learnings should be concrete and practically applicable. Never leave the responsibility for the translation of the learnings into practice fully to the individual workers. Be aware that a diploma (formal qualification) does not make somebody experienced and sufficiently competent for complex situations.

9.1.7 Golden Rule 7: Invest in People – Motivate by Participation

People are probably the most important asset of the organization: the only asset that cannot be copied by the competition. This asset is not only important for the execution of regular work but also for realize improvements in SHW and business goals. To fully use this asset, it is necessary to regularly invest in people, i.e., GR7.

Recognize your personnel's abilities, also beyond formal qualifications. When people are taken seriously, they will be more positive towards the messages from management. They will take their work more seriously, and feel responsible for addressing the associated SHW challenges.

Create regular opportunities for dialogue, and for sharing ideas and experiences. Promote good teamwork and create a community that is more than a set of individuals. Human beings are social beings. In a community, people feel responsible for each other.

Participation strengthens the feelings of ownership and makes people feel valued. It is key for strengthening intrinsic motivation, being proactive and being creative. It is essential for a commitment strategy. Any individual – manager or worker – needs regular timely and proper feedback. Feedback can be recognition or reward in case of positive achievements, but can also point out areas for (further) development or improvement. Good achievements in SHW should be recognized, socially rewarded and be conducive to further career opportunities.

People should be encouraged to suspend the work in case of acute SHW risks and to notify the management immediately. For individual workers it is a big step to suspend their work; they will not do it without a good reason. When it happens, they deserve positive feedback from management: They have done the right thing: in case of doubt, it is *better safe than sorry*. Even when in hindsight it turns out that suspending the work was not really necessary, it should be valued positively. Otherwise, nobody will ever suspend the work thereafter.

9.2 PROACTIVE LEADING INDICATORS

There is a clear need for information about actions that should be undertaken to improve SHW. To this end, the seven Golden Rules are complemented by 14 Proactive Leading Indicators (PLIs), two for each Golden Rule.[6]

9.2.1 The Challenge of Finding Good Indicators

Vision Zero implies the ambition to prevent all work-related accidents and illnesses but that does not tell you what you have to do to realize that ambition. After an accident has occurred it is relatively easy to say that it could have been prevented, but beforehand that is much more difficult. It is also easy to say that the frequency of (loss time) incidents is too high, and should be reduced, preferably to zero. Similarly, it is easy to say that something has to be improved when somebody has developed a work-related illness, e.g., a lung disease due to exposure to dust, or a burnout due to poor work organization. But in all these cases the question is: What will be effective, what should we do? There is a great need for evidence-based practical information about what should be done with predictive value for SHW performance.

> Many organizations are still relying on (lagging) indicators that reflect past performance and have no or little predictive value. Well-known examples are lost time injury frequencies, sickness absence rates, frequencies of work-related illnesses, frequencies of SHW complaints, or the number of days without accidents. Lagging indicators reflect the past and may tell you that you need to improve the SHW performance, but when the organization is committed to Vision Zero that was already clear without any metrics.

At first sight, audits of SHW management systems seem to provide the necessary information. Internal or external audits based on. e.g., the ISO 45001 standard for occupational health and safety management[7] will point out non-conformities with the standard. Solving the non-conformity leads to compliance with the standard. Though management system standards make sense for many (large) organizations, compliance to the management system standards is unfortunately not a guarantee for better SHW performance or for achieving continual improvement (see Section 8.2.4). Consequently, audits and management reviews do usually not provide the necessary information to answer the question: What should we do in order to improve SHW performance?

There is a clear need for information that points out the strengths and weaknesses of the organization's SHW preventive efforts. Information that has a predictive value for improvements and can support organizations to act more proactively to prevent accidents and work-related illnesses in the short as well as in the longer term.

9.2.2 Leading Indicators

Leading indicators are indicators that measure an exposure, factor, risk, program or control measure that occurs or exists prior to an (undesirable) SHW outcome is realized.[8] Leading indicators are therefore to be preferred over lagging indicators for decision-making and planning of SHW improvements. There are, however, various types of leading indicators: some identify existing SHW problems, and others point out strengths and weaknesses in preventive actions or SHW management. A well-known example of the first category is risk assessment. The risk assessment in itself does not prevent SHW problems but clarifies what SHW risks (potential problems) exist, and how important it is to solve them. Exposure levels, e.g., levels of noise have a similar function. An example of the second type is the percentage of work permits reviewed and found to meet the requirements.[9] The work permit example shows that leading indicators can measure preventive.

Our focus will be on *Proactive* Leading Indicators. A set of 14 PLIs was developed, two for each *Golden Rule*. The development process made use of available recent scientific evidence and evidence from good practices from industries spread over 5 continents, 23 countries and many sectors.

> The British Safety Council emphasized the lessons learned from the (construction) activities for the London Olympics in 2012. 100 million work hours thereby resulted in zero fatal accidents, historically low accident rates and good health indicators from the wellbeing programme. Key factors were visible and impactful leadership, actively using near misses, daily activity briefings at the start of work, taking action on reports or suggestions for improvement, and recognition and reward of excellent safety and health behaviour. Together with other evidence, these findings were helpful for the definition of six proactive leading indicators.[10]

Each PLI is useful for each of the three aspects SHW, and is relatively easy to measure (no additional tools required). The PLIs enhance proactivity, being early or learning and the generation of synergies.

9.2.3 The 14 Proactive Leading Indicators

Each of the seven Golden Rules is supplemented with two proactive leading indicators. However, it is often useful to start with only a few of them. Which of the seven Golden Rules deserve priority? It is then useful to start with the indicators for those Golen Rules.

Each of the indicators can be applied to safety, health and wellbeing individually, though they emphasize their commonalities and potential synergies.[11] It makes sense to first differentiate the indicators for the three aspects of SHW. That makes it easy to identify where experiences in the most developed aspect (e.g., safety) can be used for a boost in one or two of the less developed aspects (e.g., health and wellbeing). It is important to involve several internal stakeholders; dialogues and discussions in using the indicators are valuable for the improvement process. There are three options to use the indicators. They can be used in a simple checklist form (option 1), in a semi-quantitative way for estimating frequencies (option 2), or in a more extensive way for measuring them fully quantitatively (option 3).

Regarding Golden Rule 1, take leadership, demonstrate commitment, the two proactive leading indicators address respectively *visible leadership commitment* and *competent leadership.*

These indicators put the focus on two essential elements of leadership: visible and competent leadership. Through *visible leadership commitment*, leaders demonstrate their commitment to SHW and actively promote SHW improvement. It implies giving direction to SHW and business development and showing personal commitment in various situations. Expressions of commitment will be noticed by the workforce and motivate people to contribute to SHW improvement. It is necessary to involve the workforce in measuring this indicator.

Competent SHW leadership is essential to drive the development processes of Vision Zero. The SHW competence of the leaders of the organization is relatively easy to assess: are SHW qualifications and experiences part of the leaders' curriculum vitae? Did they participate successfully in programmes to develop, e.g., safety leadership or leadership in health and wellbeing? Did the organization use leadership's SHW competence as a criterion in selecting new leaders? Are there adequate leadership development programmes ongoing in the organization?

Regarding Golden Rule 2, identifying and controlling hazards and risks, the two proactive leading indicators are *evaluating risk management* and *learning from unplanned events.*

Evaluation of SHW risk management contributes to the improvement of SHW while it supports organizational learning. In many jurisdictions, it is also a legal obligation to carry out risk assessments in the production stage. Risk assessments are then regarded as the start of prevention processes: the assessments should be followed up by prioritization, an action plan and implementation thereof.

It is, however, often forgotten to evaluate whether SHW performance has been improved in practice, or not. It is also useful to evaluate *the process*, of risk assessment, action planning and implementation, as it may help to understand why some measures were more successfully implemented than others; it may also help to increase the organization's learning capacity.

Learning from unplanned events (incidents, events, cases of illness, complaints, etc.) is the second proactive leading indicator for GR 2. Learning contributes to the prevention of similar undesirable events. In safety management, it is a good practice to investigate minor incidents and near misses, especially in case these could easily have had a more dramatic impact, or in case similar incidents have happened before. Learning from unplanned events in the areas of health and wellbeing is less common, probably because health and wellbeing issues seem to be more personal, and health professionals are usually more focused on individual cases than on improvements at the system level. The recommendations after an investigation are not sufficient for better prevention. Setting priorities, making a follow-up plan, and implementing the measures in that plan are necessary to achieve improvements. Finally, an evaluation is needed to ascertain that the investigation and plan have indeed led to the intended improvements.[12]

For Golden Rule 3, defining targets, develop programmes the proactive leading indicators are *workplace and job induction* and *evaluating targeted programmes*.

The principle of proactivity is linked with being early and doing things right from the start. It is therefore a logical consequence to include attention for SHW early in the processes of hiring and selecting, and the associated workplace and job induction. Integrating SHW in induction processes demonstrates that SHW is an integral part of each job and each business process. The importance of taking care of SHW and contributing to SHW improvement should be discussed with a view on the future job: it is expected of each manager and worker to contribute, and it should be discussed what that means for the actual job, the work processes and the workplace. Workplace and job induction can be an interactive process, whereby the organization gives information on what is expected from the new manager or worker, while it is discussed whether the candidate or newly hired person is already sufficiently SHW qualified, has personal experiences with SHW, etc.

The second proactive leading indicator in this area is evaluating targeted programmes. Evaluating targeted SHW programmes (e.g., temporary campaigns) helps to verify that they are implemented as intended, and improvement goals are met. As with action plans after a risk assessment and recommendations after the investigation of unplanned events, it is important to evaluate both the result and the process of the programme. It is essential to involve the managers and workers in the programme evaluation. It is not uncommon that the various stakeholders have different experiences, perspectives and evaluations of a programme. Dialogues between the stakeholders can then trigger useful (further) improvements.

The proactive leading indicators for GR 4, *ensuring a safe and healthy system and being well organized* are *pre-work briefings* and the *planning and organization of work*.

Integrating SHW in pre-work briefings allows for the identification of context-specific hazards, risks and prevention measures prior to work. Pre-work briefings wherein attention to SHW is integrated are proven to be an effective way to prevent incidents and to prepare people to work adequately. It is essential to include some time for joint reflection before the start of the work. Is everything OK so that the work can be carried out safely and soundly? In case of doubt, what should be done to make it sure? If not, the start of the work should be suspended till everything is well prepared. The Japanese way to do this is *pointing and calling*, a group discussion whereby risks and prevention measures are identified and pointed out while communicating these findings loudly in the group. A more Western way is to give people individually the assignment to do a *last-minute risk assessment*, often using a concise checklist that is developed for each specific activity. The checklist exercise becomes easily a duty that is to be fulfilled, instead of an invitation to reflect on the work before the start.

The second proactive leading indicator for GR 4 is the *planning and organization of work*. Good planning helps to prevent unexpected situations and their associated problems. Planning and organization of work are essential for the success of every organization and for ensuring SHW. Planning is often recommended to improve safety (and quality). But planning the work processes should not be too detailed. Micro-management is often detrimental to mental health. Moreover, workers know the practical details better than those who do the planning (WaD versus WaI). The organization of work, which is all too often taken for granted, is a key factor in the prevention of psychosocial risks. The work organization determines to a high degree whether a job is only repetitive or that it comprises a variety of alternating tasks. Key factors for a good work organization are reasonable degrees of autonomy and control. That supports good teamwork and learning on the job. The work organization is indirectly important for safety too: when people have a reasonable degree of control over their work, they have the possibility to adapt the work to local or temporary circumstances (coping with variability).

The proactive leading indicators for Golden Rule 5, ensure safety and health in machines, equipment, materials and workplaces, is mostly focusing on the technological aspects, are *innovation and change*, and *SHW procurement*.

Innovation implies opportunities for improving SHW, see Sections 3.6 and 8.1.2. But it may introduce also new hazards and risks. Innovative SHW may imply a unique selling point. Organizational innovations may involve adaptation or redesign of the work organization, which is of great importance for productivity and psychosocial risks; the term workplace innovation is often used in this respect.[13]

Technological, organizational and personnel changes occur frequently in organizations and should be considered as opportunities for improvement of SHW. They can introduce new risks but also imply opportunities for SHW improvement. When change is dealt with as early as possible, it may significantly improve SHW. Management of change is a regular issue in safety management systems. But it is all too often limited to making sure that when technological items are replaced, the new items are as good as the original ones. But why not search for items that perform better? Organizational and personnel changes can have an impact on the prevention arrangement in an organization but are usually more difficult to assess than technological changes.

The second proactive leading indicator in this area is *SHW procurement*. Procurement can determine SHW risks for a long period, especially in the case of expensive equipment. The indicator triggers the systematic use of procurement for SHW improvement. Specifications that are relevant for SHW should be included in tenders or invitations to vendors, for products as well as for services. There may be a dilemma: to specify the SHW in detail gives certainty about the SHW qualities of the items procured. But especially for more complex items, it can be better to specify the requirements not in-depth, but ask the potential vendors to come up with their best solutions. In that way, SHW innovation is promoted, and the vendors may come up with innovative solutions that are beyond the client's expectations.

The indicators for Golden Rule 6, improving qualifications and the development of competences, are *initial training and education* as well as *periodic refresher training*.

Initial SHW training and education ensure that people are well-qualified for the SHW aspects before they start the job. This is important for (top) managers as well as for workers. It is important that basic SHW education is becoming already integrated into the regular educational systems, e.g., for managers in business schools and technological universities and for workers in all kinds of vocational education and training.[14] For technology-related jobs, it is increasingly a trend that companies cooperate with educational institutes and offer internships. Employers sometimes hesitate to invest in education and training of temporary workers. It is then essential that workers are already well prepared for dealing with SHW risks in a variety of work settings.

Refresher training ensures that leaders' and workers' knowledge and skills on SHW remain up to date. People tend to forget things and to pay less attention to things that seem not so relevant in everyday practice. That is why refresher training on critical SHW aspects is important. In areas where the state of the art is rapidly developing, such as information technology, refresher training should also be used to ensure people master the latest developments relevant to their work.

For Golden Rule 7, investing in people and motivating by participation, the two indicators are *suggestions for improvement* and *recognition and reward.*

Suggestions for improvement are important because the workforce has tacit knowledge based on their practical experience. Suggestions for improvements should be welcomed and taken seriously. Bottom-up suggestions can lead to practical solutions. Do not wait passively till somebody comes up with a suggestion in an idea box. Suggestions may come in the form of unanticipated questions, or proposals for solutions. The process of generating suggestions can be organized by using SHW circles that are supported by a process owner[15] or by participative brainstorm meetings to generate alternative solutions for concrete SHW problems.

People will only be willing to share their ideas when these are taken seriously, which presupposes that they receive timely and adequate feedback. Not all suggestions will be useful and followed up; in case a suggestion is not adopted, people should receive an explanation as to why their suggestions are not taken up. People should anyhow be thanked for their suggestions and for their contribution. In this way, active commitment is stimulated, a key element of a strong prevention culture and a commitment strategy.

Recognition and reward are examples of positive feedback and express that people and their efforts matter. Recognition and reward for SHW involve showing appreciation for engaging in desired SHW behaviours including coming up with initiatives or suggestions for improvement. Social rewards and recognition are often much more important than financial rewards. Commitment to improve work and business processes should be appreciated independent of the improvements realized. It is key that commitment to SHW should be positive for career opportunities in the organization.

9.2.4 The Complementarity of the Proactive Leading Indicators to ISO 45001

The seven Golden Rules include several aspects of sound traditional OSH management, and there is a logical coherence between several ISO 45001 requirements and some Proactive Leading Indicators.

ISO-type management systems such as ISO 45001[16] are strongly focussed on systematic management of SHW risks and opportunities. Though the formal aim of the standards is to ensure a process of continual improvement, in practice it may develop into a system to cope with all changes without much improvements (see Section 8.2.4). But that need not be.

ISO-type management systems can develop a much stronger focus innovation, by using proactive leading indicators. The indicators can put additional focus on the early stages of development and design and can strengthen processes of organizational learning.

Some ISO requirements correspond to proactive leading indicators but have a more limited and reactive focus. The clauses in ISO 45001 for dealing with change and procurement focus respectively on mitigation of adverse effects and on ensuring conformity. The focus of indicator 5.1 for Innovation and change is on using innovation and change as an opportunity for SHW improvement. Similarly, indicator 5.2 on procurement aims at using procurement as an opportunity for improvement.

Four proactive leading indicators do not correspond with ISO 45001 requirements. This is the case for indicator 1.2 competent leadership, 4.1 pre-work briefings, and 7.2 for recognition and reward. While workplace and job induction (indicator 3.1) are only mentioned in the informative annex of ISO 45001 and are not an ISO requirement. Compliance with ISO 45001 does therefore not ensure that these opportunities for improvement are acknowledged and systematically managed.

The added value of the other indicators compared to ISO 45001 lies in the options for frequency estimation or quantitative measurement. By estimating or determining the relevant frequencies or percentages of organizations, the stakeholders involved can better evaluate whether they are satisfied or not with the outcomes.

The set of proactive leading indicators was developed to make key aspects of SHW prevention efforts measurable, to stimulate SHW improvement through innovation and to support the management of continual improvement. Proactive leading indicators are complementary to the ISO 45000 series and can help certified companies to strengthen their processes of continual improvement. The use of the Proactive leading indicators generates important data that can serve as input for the management review.[17] The indicators also offer opportunities for benchmarking, internally or externally.

9.3 VICIOUS AND VIRTUOUS PROCESSES

While the seven Golden Rules and the Proactive Leading Indicators were developed specifically for Vision Zero, there are several other not-so-well-known or new approaches that can significantly contribute to the process of long-term continual improvement.

The first we propose is using self-reinforcing processes positively. In the case of the ill-famed vicious process, results go from bad to worse. But in the case of the less well-known virtuous processes, the results go from promising to excellent. These self-enhancing processes continue for longer periods. Both vicious and virtuous processes have a circular character; they cannot be understood through linear thinking. To better understand the principle, let us consider vicious cycles first.

9.3.1 Vicious Processes and How to Stop Them

A vicious circle does not stop easily and has the tendency to continue itself. It will stop when one of the steps in the process is blocked or when the conditions are changed significantly. A well-known example on the individual level is: You have little self-confidence and are uncertain about the capacities for a certain task, that is part of your job. So, after some hesitation you do it, but unfortunately the results are (as you already feared) not brilliant. In other words, you feel that you have failed and will be criticized for it. This further affects your self-confidence and increases the uncertainty about your capacities. This negative process can be stopped by social support and more positive feedback and recognition.

The examples in Table 9.1 illustrate that vicious processes are often responsible for persistent SHW problems.

When vicious processes are recognized, it works best to analyse each step of the cycle. The circumstances may contribute to the logic of the respective steps. In theory, only one step has to be blocked to stop the vicious process. But it often makes sense to search for ways to discourage several steps.

9.3.2 Initiating Virtuous Processes

Fortunately, it is possible to make use of positive, self-reinforcing processes, or virtuous processes with increasingly positive outcomes. Again, it is often a combination of the steps in the process and the circumstances that make the process work. Creating the conditions for virtuous processes and initiating them is an interesting but often forgotten way to come to long-term continual improvements in SHW. First an example: You are challenged to do a difficult task and you have sufficient self-confidence. You do it and you receive compliments for the results, even though the results were not perfect. You get a positive feeling about the process, the challenge and the outcome. Your self-confidence is growing further, etc. Note that this virtuous process mirrors the vicious process described in former section. A similar virtuous process may refer to competence instead of self-confidence (see Figure 9.2).

Besides the individual level, virtuous processes can also take place on the organizational level. This implies the possibility to create processes and conditions that

TABLE 9.1
Three examples of relevant vicious processes and possibilities to stop them

Area	Examples of Vicious Processes	What Can Be Done to Stop It?
Compliance triggers bureaucracy	The organization's policy focuses on compliance → In it is important to show to the authorities that measures are being taken → Paper work becomes very important → Paper safety becomes a goal in itself → When problems turn up, papers are the most important *evidence* of compliance → The focus on compliance is intensified.	The process can be internally blocked by regularly discussing the relevance of procedures for SHW in practice. It can be blocked externally by authorities that do not accept paper proof of compliance.
Micro-management frustrates professionals	Manager does not trust the capacities of some professionals → Manager wants more control and starts to micro-manage → Professionals become frustrated and stressed → Professionals deliver fewer promising results and show symptoms of severe stress → Manager's trust in the professionals is further diminished.	This process can be stopped by an open conversation between the manager and the professionals involved. Or by a colleague-manager who gives encouraging feedback to the manager.
Sickness absence and work pressure	Long-term sickness absence is high and increasing → more work to be done with fewer people → greater burden of work → sickness absence increases further.	Recurrence of short-term sickness absence is a warning for long-term absence. Organize temporary personnel with sufficient experience; or consider changes in the work organization to give people more possibilities for control over their work processes.

have a long-lasting and increasingly positive organizational impact. Such processes can be useful to generate business benefits through good SHW, to strengthen SHW leadership, to develop shared values and creating trust, to promote positive SHW behaviour and to encourage intrinsic motivation for SHW, etc.

A virtuous process can also explain the power of exemplary behaviour by (formal or informal) leaders, see Figure 9.3. The virtuous process depicted in Figure 9.3 is useful for creating a more proactive culture. Give room to undertake initiatives to some SHW champions who are already intrinsically motivated to improve SHW. They will combine the improvements with their enthusiasm, which is likely to inspire also others.

There are more options for virtuous processes that improve SHW. Some examples are given in Table 9.2.

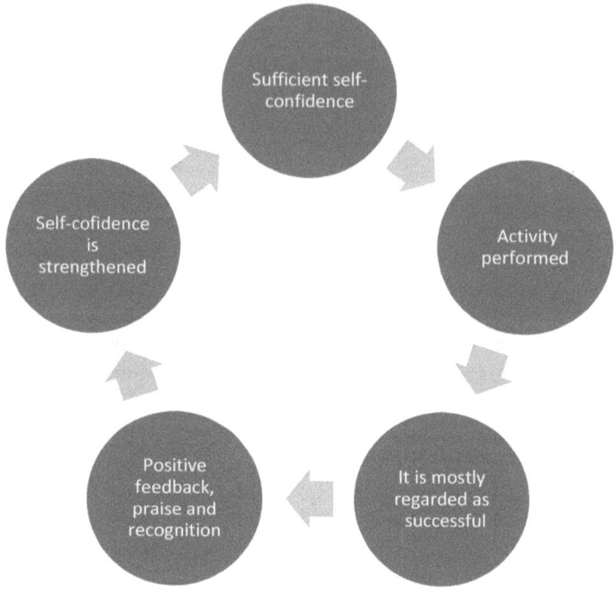

FIGURE 9.2 An example of a virtuous process at the individual level.

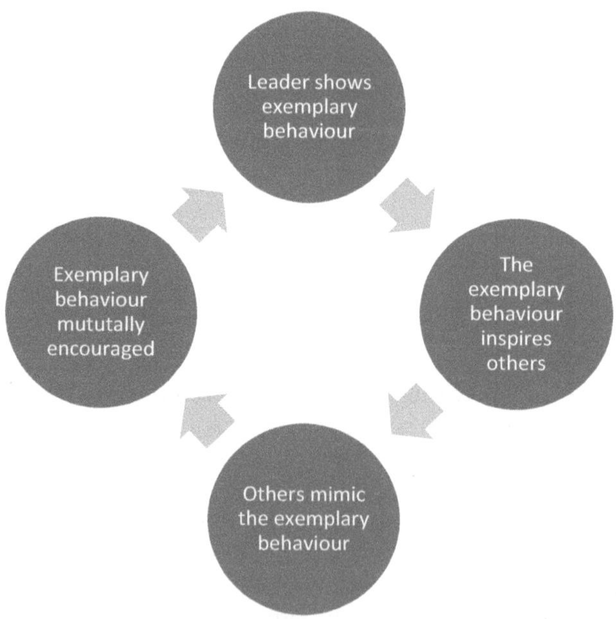

FIGURE 9.3 The virtuous process associated with exemplary leader behaviour.

TABLE 9.2
Examples of virtuous processes for improving SHW

Area	Examples of virtuous processes	What can be done to initiate it?
Leadership and team work	Manager sees team performance as key → manager encourages team work → team members improve cooperation → Team members are proud of having improved team performance → manager is recognized as a good team leader → manager further encourages team work.	Manager realizes that team performance is more important than individual performance.
Experience and professionality	Experienced professionals have developed personal mastery over their work processes → They recognize existing and upcoming risks and opportunities for improvement during their work → They are able to use opportunities and to minimize risks → That makes them even better professionals.	Give experienced professionals room to organize their work and use their ideas to improve work and SHW.
Good health and wellbeing and productivity	The organization invests in good health and wellbeing → High productivity at work and increased creativity → High business competitiveness → Investing in worker health and wellbeing recognized as key factor for good business results.	Start a programme to invest in health and wellbeing for all of the personnel (not only for those at risk).
Purpose and intrinsic motivation	People see a possibility to contribute during their work to their personal purpose → They try it out → This work is rewarding → People are increasingly intrinsically motivated for their work as it contributes to their purpose.	Align the organization's purpose with that of the people.

9.4 DEALING WITH WEAK SIGNALS AND EARLY WARNINGS

Accidents and illness are often preceded by weak signals or early warnings that can help to anticipate and prevent them. Those signals should be recognized and taken seriously. The interpretation of weak signals is difficult, as the signals can be of an unusual nature, appear unexpected, and are often ambiguous, while the people involved are usually unprepared. The principle of High Reliability Organizations (HROs) *reluctance to simplify* is then important. Weak signals should be considered from multiple perspectives, i.e., *allowing ambiguities*. People on the shop floor may have valuable professional experience, while experts may have predominantly theoretical knowledge that may be of help.

The human capacities for recognition and evaluation of signals are at the centre of concerns for high-reliable organizing which is relevant for safety in complex situations. Though IT systems with sensors and artificial intelligence play a growing role.

Well-developed capacities for recognition are a key characteristic of a proactive prevention culture and are often called *organizational mindfulness.*[18]

The alertness is not only a matter of ratio and knowledge. Intuition or gut feeling is important as well.[19] Intuition is not really understood scientifically but seems to be based on observations that are stored in the mind, unconsciously or half consciously. These former observations play a role in human memory. In a way, intuition builds on earlier experience and it can be regarded an expression of professionality.[20] Memories play a role in the interpretation of what seems to be side-observations; in HRO literature this is called *sensitivity to operations.* It may come in the form of an *uneasy feeling*, or a soft sense of alarm. Intuition is never a direct way to know the truth, but a warning that should be taken seriously; a reason for further investigation. Roggeveen states in relation to safety:

> The signals may appear as feelings, such as being surprised, puzzled or anxious about something not looking, sounding, smelling as it normally does. Trust those feelings! They are a solid clue that something is really wrong. Do not ignore them but investigate their origin, before continuing your planned course of action.[21]

To support the recognition of weak signals, the Vision Zero surprise check presented in Table 9.3 may be of help.

Weak signals may also concern individual people. They may have problems in meeting the demands of work. They may have problems with the (toxic?) behaviour of leaders, colleagues or clients, or have problems at home which keep them awake at night. They may have some health issues that limit their capacity for work. Perhaps they have financial problems. They may worry about problems with their adolescent son. They may lack self-confidence for a difficult task, etc.

Someone has to have the courage to address the weak signals. To speak up and ask questions like: What does this mean? What is going on here? Are we sure that this is going right? When a weak signal is recognized and reported, the first response may be to ignore it as irrelevant (e.g., This is a negligible phenomenon; I am sure there is nothing wrong). That would be jumping to conclusions and discourage people from speaking up in case of weak signals. There may not (yet) be a rational explanation, nor a ratio that the weak signal is an early warning or not. Leaders should appreciate the fact that somebody has the courage to speak up and to acknowledge that weak signals can sometimes be early warnings.

Organizational change or digitalization in organizations often implies subtle but significant changes that may trigger uneasy feelings. Organizational change may introduce new roles which are not yet fully clear. It may imply that existing roles become obsolete, which affect the people involved. What does that mean for mutual trust and shared values? What consequences have changes in rights and accountabilities? How does ongoing digitalization influence the organizational culture? What are the unexpected impacts of frequently working from home? Of less face-to-face communication? Does it lead to less team feeling? Does it lead to privacy issues? Further

TABLE 9.3
The Vision Zero surprise check

Human signals	Not everybody in the team knows the people they have to work with
	Somebody in the team feels not really fit for work
	High personnel turnover or structural understaffing
	Everybody is always very busy
	Complaints about the toxic behaviour of leaders or colleagues
	Problems notified by a whistle blower
	Unclear messages received by a confident
	People having repetitive complaints (e.g., back pain)
	Worker health complaints
	Signs of worker fatigue
	Individuals with repetitive short-term sickness absence
	Inconsistent or incorrect use of safety devices or personal protective equipment
	Somebody has an uneasy feeling or a soft sense of alarm
Technological signals	Signs of physical damage
	Problems with technical components occurring before the expected end of their life cycle
	Radical innovations using new materials with partially unknown physical properties
	Corrosion
	Strange sounds
	Unexpected high or low temperatures
	Unexpected vibrations
	Unexpected changes in pressure
	Unexpected odours
Organizational signals	Frequent changes in ownership or senior leadership
	Frequent changes in priorities; a zigzag policy
	Key stakeholders with conflicting interests or value systems
	Structural overwork
	Repetitive problems that are difficult to explain and solve
	Substandard house-keeping
	On-going cost-cutting
	SHW budget reductions
	Management and workers see each other as 'us versus them'
	Conflicts between leaders or departments
	High (frequency of short-term) sickness absence

Source: Part of the Vision Zero Surprise check is elaborating on CCPS, (2012) *Recognising Catastrophic Incident Warning Signs in the Process Industries*. Center for Chemical Process Safety. Hoboken, NJ: John Wiley & Sons.

individualization of workers? What impact does Artificial Intelligence perform on people in the organization? Etc.

In all these cases it is necessary to take the symptoms seriously. To start open conversations and listen carefully to what people say, and find out what are the underlying factors of the problem signalled. This requires modesty from the side of the leaders, as well as emotional intelligence. Even when the problems are personal, the organization may be able to help solving or mitigating the personal problems. From a business perspective that can be beneficial for the productivity of the person involved. Taking such issues seriously can greatly contribute to the valuation by the personnel of the organization as an employer of choice.

9.4.1 Weak Signals, Complexity and Emergence

Complexity characterizes the behaviour of a system whose components interact in multiple ways leading to non-linear and seemingly random phenomena. *Emergence* thereby refers to properties or behaviours of the system that cannot be explained by the properties of individual components but only by interactions of multiple components of the system. As emergence is not a controlled process, the resulting risks require alertness on weak signals and potential early warnings. Organizations, however, are man-made, and complexity in organizations is the (implicit) result of choices made. When other choices are made, complexity and emergence might be reduced or even disappear (compare with the principle of simplification in Table 6.2).

Human decisions may cause complexity in an organization. When this leads to the emergence of new unexpected SHW risks, that is sometimes seen as inevitable. But that resembles the old habit of accepting accidents as *fate* or as *acts of God* from more than a century ago. The concept of complexity is sometimes used to cover up an attitude of complacency. Other choices in the earlier stages of development and design would perhaps have prevented the unexpected SHW risks from emerging.

9.5 USING THE PRACTICAL WISDOM OF THE WORKFORCE

Risk awareness can improve through worker involvement or when employee representatives are engaged in SHW activities.[22] The so-called new view on safety has caused renewed attention to the practical wisdom of workers. The tacit knowledge of workers and its relevance for SHW is rediscovered in response to the excessive attention to human error.[23] This is most clearly explained by Hollnagel as what he called Theory W compared to Theory Z, see Table 9.4. Hollnagel refers thereby mainly to the capacities of people to deal with disturbances and variability.

In more practical terms, there is a growing realization of the value of Work as Done (WaD) versus Work as Imagined (WaI), and the potential gaps between the two. The underlying idea is that deviations from formal prescriptions are often based on the practical wisdom of the workers involved. The gap between WaD and WaI should therefore be regarded as an opportunity to learn instead of problematic by definition.

Instead of detailed prescriptions or standardization of tasks, it is better to give workers, individually or on team level, a reasonable degree of control over their work.

TABLE 9.4
Two mental models on the capacities of people to deal with disturbances and variability

Theory W	Theory Z
Systems are well-designed and scrupulously maintained	People learn to identify and overcome design flaws and functional glitches
The procedures that are provided are complete and correct	People can recognize the actual demands and adapt their performance accordingly
People behave as they are expected to and have been taught or trained to	When procedures must be applied, people can interpret and apply them to match the conditions
System designers have been able to foresee and anticipate every contingency	People can detect and correct when something goes wrong or when it is about to go wrong, and hence intervene before the situation seriously worsens

Source: Hollnagel, E., (2009) Extending the scope of the human factor, in: *Safer Complex Industrial Environments – Human Factors Approach*. Boca Raton: CRC Press. pp. 37–60.

Then they can take appropriate action when needed. The work organization is a key factor in this respect. Of course, there will always remain issues that are beyond the control of the individual workers or the team, where the involvement of a supervisor or higher manager is needed.

Creating a work organization that allows for sufficient control over the work processes, and for learning by doing is the more structural way to use and develop the practical wisdom of the workers. But some organizations prefer to tap the wisdom of the workforce without empowering them. Then it is important for the leaders, and also for SHW experts, to engage in meaningful dialogues with workers, whereby listening and asking open questions is more important than telling and explaining, let alone inspecting. The advantage of such conversations is that they increase the social capital of the workforce and help leaders and workers to jointly see *the bigger picture* and to weigh interests and possibilities for improvement.[24]

9.6 STARTING WITH VISION ZERO AFTER A FATAL OR SERIOUS SHW EVENT

Organizations can commit themselves to Vision Zero for a variety of reasons: Because they want to become an attractive employer or a frontrunner in SHW, or because they see it as a social responsibility or as a way to improve the image and profitability of the organization. Perhaps the CEO is personally motivated to prevent accidents and ill health. These reasons arise from internal considerations. But there is also a category that is triggered by fatal or serious accidents or by cases of serious occupational illness (e.g., occupational cancers or work-related suicide).

Such events imply human misery as well as serious problems for the organization and its leaders. It often raises questions like: How could this happen? Why weren't we able to prevent this? It also triggers normative statements like: This may never happen again! It may cause a crisis and the need for self-reflection. A crisis that should not be wasted, but used as an opportunity for improvement. A moment to (re)discover the value of human life and health, to redefine the purpose of the organization, and to adopt Vision Zero; see also Section 5.1.

The first activity is to organize a good and thorough investigation to answer the questions: How did this happen? What can we learn from this terrible event? Take the initiative for such an investigation with the only aim to learn (authorities may also investigate the event but their main aim is usually to reconstruct what happened and to identify non-compliances and liabilities). Involve one or a few external experts in the investigation besides your own people. Make sure the investigation is not focused on identifying a human error as a root cause or finding a scapegoat. Ensure that the investigation identifies missed opportunities for leadership and workers to prevent the event from happening. Consider technological, organizational and human factors. Address underlying factors. Perhaps there were blind spots or taboos that impeded the recognition of risks and warning signals. Invite people to address these. Identify and communicate lessons learned and discuss them with a range of internal (and external) stakeholders and make sure the lessons learned are followed up with an action plan and implementation of the lessons learned.

Communicate that implementing the lessons learned is only a first step. That it is important to invest in better prevention. Vision Zero requires a long-term approach. The development of a strong prevention culture will require at least five years. Develop a framework for three or five years of strengthening prevention that addresses technological, organizational and human factors. Make a SMART plan for the first year. Define aims and make sure there are milestones and evaluations. Strengthen the SHW expertise in the organization, in the line organization and dedicated SHW staff. Involve the workforce, also via worker representatives. State and accept consequences for design, new activities and purchasing (including working with contractors). Do not limit the programme only to safety, or only to health, but address the three aspects of SHW, generating synergies.

To assess weaknesses in SHW prevention it is advisable to make use of the seven Golden Rules.[25] and the Proactive Leading Indicators.[26] Repeat that after some time, to monitor progress. Make sure successes are realized and widely communicated without making it a good news show. Also communicate the challenges and difficulties encountered, and the plan to overcome them (after all Vision Zero is a long-term challenge). Consider the possibility of introducing an annual SHW day, wherein prevention and safeguarding life and health in the organization are central.

9.7 THE DON'TS OF VISION ZERO

For every serious ambition, you have to do a lot to realize it. But the other side of the coin is that there are also a number of things that you should refrain from doing. Top sporters refrain from partying, from drinking beers at night with their friends, etc. The don'ts of a strong ambition are often just as difficult to realize as the dos.

Two types of don'ts are important for Vision Zero: don'ts to avoid doing the wrong things, and don'ts to ensure that things should not be overdone. It might be useful to consider the 12 design principles for Worker Friendly Design as set out in Section 6.5.5. Most of them imply do's but implicitly also some don'ts.

9.7.1 Examples of Avoiding Doing the Wrong Things

- Don't make people, departments or units accountable for zero as a target
- Don't use Vision Zero as a slogan only, without concrete activities to improve SHW and open two-way communication
- Don't focus SHW management only on the stage of production and maintenance
- In case of doubt, don't take a SHW risk
- Don't take the existing situation and its associated risks for granted
- Don't accept SHW silos; instead create synergies by working together or another organization of work
- Don't use (financial) rewards to motivate people; it will only have a short-term impact and destroy their intrinsic motivation
- Don't create or accept role conflicts or hassles.

9.7.2 Examples of Too Much of a Good Thing

- Don't make too many rules and prescriptions: It will not work and in the end be counterproductive.
- Don't control people or their work intensively: It will affect their dignity and wellbeing, and discourage ownership.
- Don't create or accept a disbalance between work and private life: It will affect human life and health of the workers and their families; it may also trigger talented people to look for another job elsewhere.
- Don't accept exceptionally long working hours in your organization: working structurally more than 55 hours per week is globally the main cause of work-related death, see Chapter 1.
- Don't focus too much on making profit or on making career: At the end of the day, wellbeing and happiness are more important.
- Don't plan the available time full of productive activities: allow time for reflection, creativity or meditation (such *spare time* is useful to avoid a tunnel vision, for reflection, and for out-of-the-box thinking).
- Don't trust too much on your planning: That may imply loss of flexibility; Conversations about updating or adapting planning are just as important.
- Don't expect people to be top performing all the time: Work is not top sport; allow time for physical and mental recovery, especially after a stressful period
- Don't fully plan the activities of others: Give people room for autonomy to arrange their work for themselves
- Don't eliminate all technological or organizational redundancies: They can be useful as buffer or back-up

9.8 CONCLUSIONS

The seven Golden Rules and the 14 Proactive Leading Indicators were specifically developed for the implementation of Vision Zero. The Golden Rules build on good practices with a clear focus on prevention. They combine traditional and innovative perspectives while offering flexibility for tailor-made approaches. They do not prescribe activities but offer guidance to the companies and their leaders. The decision to adopt the seven Golden Rules should be made by top management as part of their commitment to Vision Zero. The Proactive Leading Indicators support the Golden Rules and focus on factors to boost proactive prevention or the development of a prevention culture. Each indicator can be used in a simple or a more extensive way. The indicators can help translate good practices in one area to other less developed areas. The proactive leading indicators are complementary to ISO 45001 management systems and can help certified companies develop a stronger focus on innovation and prevention.

The introduction of virtuous processes is a way to introduce self-sustaining processes of long-term improvement. Like vicious cycles, they do not fit in linear cause-and-effect thinking. In a circular process, issues are both cause and effect. Their self-enhancing nature makes they have the tendency to go on for longer periods with increasing impacts. Initiating virtuous processes can therefore have a long-lasting positive impact.

To recognize and evaluate weak signals as potential early warnings and deal adequately with them is vital for prevention. Weak signals should be taken seriously, though they often have an ambiguous character. Human, technological or organizational signals may precede accidents or illnesses and imply the opportunity to take preventive action before negative effects materialize. The Vision Zero surprise check may help in this respect.

Using the practical wisdom of the workers is another strategy for improvement. In practical terms, there is a growing realization of the value of Work as Done and Work as Imagined. A gap between the two should be regarded as an opportunity for organizational learning.

A fatal or serious SHW event often implies a crisis and the need to reconsider existing SHW strategies. It is a good moment to adopt Vision Zero. Tips for the process of implementing Vision Zero in such cases are given.

Finally, for Vision Zero there are also a number of don'ts, of things to avoid. These are either examples of doing the wrong things or of good things that should not be exaggerated.

NOTES

1 ISSA, (2017) *The Seven Golden Rules.* Geneva: International Social Security Association. https://visionzero.global/all-documents?title=&field_category_target_id=458&field_publication_date_value= (Accessed 19 December 2024).

2 ISSA, (2017) *The Seven Golden Rules.* Geneva: International Social Security Association. https://visionzero.global/all-documents?title=&field_category_target_id=458&field_publication_date_value= (Accessed 19 December 2024).

3 Besides the ISSA guide already mentioned, there is also the guide 7 Golden Rules for small businesses; there are also a few sector versions, e.g., for the construction sector.

4 Leveson, N., (2020) *Safety III – A Systems Approach to Safety and Resilience*. Cambridge: MIT.

5 Dyreborg, J., Johnstone Lipscomb, H., Nielsen, K., Törner, M., Rasmussen, K., Frydendall, K.B., Bay, H., Gensby, U., Bengtsen, E., Guldenmund, F., Kines, P., (2022) Safety interventions for the prevention of accidents at work: A systematic review. *Campbell Systematic Reviews*, e1234. Doi: 10.1002/cl2.1234.

6 ISSA, (2020) *Vison Zero – Proactive Leading Indicators – A Guide to Measure and Manage Safety, Health and Wellbeing at Work*. International Social Security Association. https://visionzero.global/all-documents?title=&field_category_target_id=458&field_publication_date_value= (Accessed 19 December 2024).

7 ISO, (2018) *Occupational Health and Safety Management Systems – Requirements with Guidance for Use*. Geneva: International Organization for Standardization.

8 AIHA, CSHS, (2020) *Best Practice Guide for Leading Health Metrics in Occupational Health and Safety Programs*. Falls Church (VA): American Industrial Hygiene Association and Center for Safety and Health Sustainability.

9 See AIHA, CSHS, (2020) *Best Practice Guide for Leading Health Metrics in Occupational Health and Safety Programs*. Falls Church (VA): American Industrial Hygiene Association and Center for Safety and Health Sustainability, for more examples.

10 The proactive leading indicators on visible and competent leadership (1.1, 1.2), learning from unplanned events (2.2), pre-work briefings (4.1) and on suggestions for improvement (7.1) and recognition and reward (7.2) are partly based on the evaluations of the London Olympics 2012.

11 Zwetsloot G.I.J.M., Leka S., Kines P., Jain A., (2020) Vision Zero: Developing proactive leading indicators for safety, health and wellbeing at work. *Safety Science*, 130, 104890. Doi: 10.1016/j.ssci.2020.104890.

12 Compare with Drupsteen, L., Groeneweg, J., Zwetsloot, G.I.J.M., (2013) Identifying critical steps in learning from incidents, *Journal of Occupational Safety and Ergonomics*, 19 (1), 63–77. Doi: 10.1080/10803548.2013.11076966.

13 Oeij, P.R.A., Rus, D., Pot, F.D., (2017). *Workplace Innovation – Theory, Research and Practice*. Cham, Switzerland: Springer.

14 Bollmann, U., Boustras, G., (eds) (2021) *Safety and Health Competence – A Guide for Cultures of Prevention*. Boca Raton: CRC Press.

15 Schoër, A., Sochert, R., (2000) *Health Promotion Circles*. Essen: BKK.

16 ISO, (2018) *Occupational Health and Safety Management Systems – Requirements with Guidance for Use*. ISO 45001:2018. Geneva: International Standardization Organization.

17 ISO, (2018) *Occupational Health and Safety Management Systems – Requirements with Guidance for Use*. ISO 45001:2018. Geneva: International Standardization Organization, Clause 9.3.

18 Weick, K.E., Sutcliffe, K., (2007) *Managing the Unexpected*. San Francisco: Jossey-Bass.

19 De Bruine, H., (2018) Gebruik het gevoel van ongemak! Een onderzoek naar het bevorderen van hoog betrouwbaar organiseren (Use your uneasy feelings! An investigation into the promotion of high reliability organising). Doctoral thesis. GVO drukkers & vormgevers B.V. | Ponsen & Looijen.

20 Kahneman, D., Klein, G., (2009) Conditions for intuitive expertise, *American Psychologist*, 64 (6), 515–526. Doi: 10.1037/a0016755.

21 Roggeveen, V., (2022) *The influence of leadership on the prevention of safety incidents: on risk reduction, leadership, safety principles and practices*. Doctoral thesis. University of Leiden (NL). p. 254.

22 EU OSHA, (2019) *Third European Survey of Enterprises on New and Emerging Risks* (ESENER 2019), Overview Report of how European workplaces manage safety and health. Bilbao: EU OSHA. https://osha.europa.eu/en/publications/esener-2019-overview-report-how-european-workplaces-manage-safety-and-health (Accessed 26 February 2024).

23 Podgorski, D., (2010) The use of tacit knowledge in occupational safety and health management system, *International Journal of Occupational Safety and Ergonomics*, 16 (3), 283–310. Doi: 10.1080/10803548.2010.11076845.

24 IOGP, (2022) *Learning from Normal Work*. London: International Association of Oil and Gas Producers. Report 642.

25 There are two ISSA guides for the seven Golden Rules, of which one for small organizations. All ISSA Vision Zero guides can be downloaded free of charge from https://visionzero.global/all-documents?title=&field_category_target_id=458&field_publication_date_value=

26 ISSA, (2020) *Vison Zero Proactive Leading Indicators – A Guide to Measure and Manage Safety, Health and Wellbeing at Work.* International Social Security Association. https://visionzero.global/all-documents?title=&field_category_target_id=458&field_publication_date_value= (Accessed 19 December 2024).

10 Broader Perspectives

In this chapter we focus on broader perspectives, presenting the meaning of Vision Zero beyond SHW or beyond the organizational level. We start with broader synergies.

10.1 BROADER SYNERGIES

Synergies imply opportunities for mutual improvement. Synergies can be defined as a process of interaction that would provide benefits above and beyond those provided by the units separately.[1] It is a process of mutually positive interactions. Striving to realize synergies is at the heart of Vision Zero. That is most prominent for the synergies between safety, health and wellbeing, and between SHW and good business. But there are more opportunities for synergies. These will be concisely sketched here.

10.1.1 Synergies between Vision Zero and Quality Management

In organization practice, SHW is very often related to managing quality. The Zero Defects movement was already mentioned as one of the three roots of Vision Zero. The principle of *doing things right* from the start and of *doing the right things* also stems from the area of quality management. In Chapter 2, it was mentioned that many safety and quality tools are the same, and for both areas it is important to prevent deviations from normal. The same or very similar techniques can also be used for promoting or enabling improvements for instance by using Quality Circles or Health Promotion Circles.[2]

OSH management systems and quality management systems are often based on ISO standards (the ISO 45000 series and the 9000 series) and have many aspects in common. Since 2014, ISO has harmonized requirements for all ISO management systems to facilitate their integration and synergies. Many organizations have integrated staff functions for SHW and quality management, often also combined with environmental management. The common aspects of the management systems, and of tools and approaches offer opportunities for synergies between SHW and quality.

The added value of quality management for the business is often easier recognized than for SHW, as higher quality usually means a higher price and more profit. Quality

DOI: 10.1201/9781003621225-10

management therefore tends to have a higher status than SHW in many organizations. Making use of the strengths of quality management therefore facilitates SHW management. On the other hand, the experiences with developing a prevention culture and creating ownership on the shop floor are simpler developed for Vision Zero, as each worker has a personal interest in SHW.

For Total Quality Management, it is important that a high quality of products and services is generated through high-quality production processes, supported by other high-quality processes. The excellence of the technologies used, the workplaces, the work environment and the people should then be considered. TQM therefore also comprises the quality of work. That implies, again, opportunities for synergies between Vision Zero and TQM.

The idea of synergies between SHW and Quality is most prominent in the concept of *Operational Excellence* and the related concept of World Class Manufacturing.[3] Well-known examples of approaches to Operational Excellence are Lean Manufacturing, Six Sigma and Kai Zen.[4] These approaches focus on involving workers, using their practical knowledge for the improvement of production processes and creating a prevention-oriented culture. In World Class Manufacturing, there is a greater emphasis on using innovative technologies and Industry 4.0 or 5.0 for promoting Operational Excellence. Many of the elements of Vision Zero, its philosophy and strategies for early effective prevention are also important elements of Operational Excellence and World Class Manufacturing, and vice versa.

The Japanese electronics industry visualized its Vision of their future as world-class excellence through 5-Zero Manufacturing (see Figure 10.1). That concept comprises zero defects, zero production losses, zero late delivery, zero accidents and zero downtime. The 5 Zeros are clearly interdependent. Any accidents imply downtime and production losses. Late delivery implies greater pressure to speed up, increasing the likelihood of errors and accidents, etc. Only by seeing 'the bigger picture' and taking into account the 5 Zeros and their synergies, industry can become really excellent. Experiences in these approaches may therefore strengthen strategies and prevention for Vision Zero and vice versa.

Though the goals of 5-Zero manufacturing as depicted are all based on a business rationality and require ratio to achieve them, in practice that will always be associated with a prevention-oriented culture. Each of the 5-Zero goals requires visible leadership commitment, competence and a culture of participation and learning wherein leaders, and the overwhelming majority of workforce are pursuing the five goals proactively. The Proactive Leading Indicators presented in Section 9.2 can rather easily translated into proactive leading indicators for Total Quality Management or World Class Manufacturing. That would also broaden the support in the organization for the Proactive Leading Indicators for Vision Zero.

There are striking similarities between the European Foundation for Quality Management's (EFQM) latest model for business excellence and Vision Zero. See Figure 10.2.

The EFQM model comprises three main areas: (1) *Direction* which refers to two criteria: (a) Purpose, vision and strategy, and (b) Organizational culture and leadership; (2) *Execution* which comprises three criteria; (c) Engaging stakeholders,

FIGURE 10.1 The 5-Zero Manufacturing Vision of the Nippon Electric Control Equipment Industries Association (NECA). (From NECA, (2017) Towards 'Future Vision 2025', Manufacturing for the future proposed by NECA, in: *Profile of NECA*. Tokyo: Nippon Electronic Control Equipment Industries Association. p. 6.)

(d) Creating sustainable value; and (e) Driving performance and Transformation and (3) *Results*, with the criteria (f) Stakeholder perceptions and (g) Strategic and operational performances. The three main areas as well as the seven criteria could have been written with Vision Zero in mind.

10.1.2 Synergies between Vision Zero, Sustainability and Corporate Social Responsibility

Vision Zero is in several ways related to or similar to the quest for sustainability. There is a widely growing awareness that economy and ecology are closely connected and that ultimately a sustainable economy cannot exist without

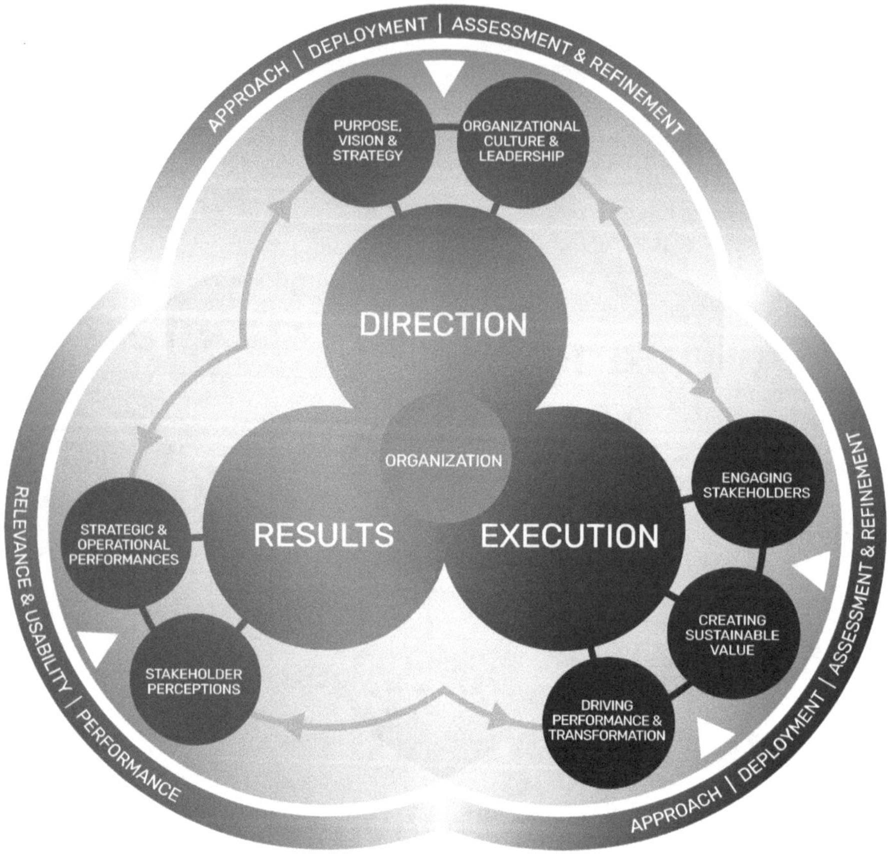

FIGURE 10.2 The EFQM Model 2025 of business excellence. (From EFQM, (2024) *The EFQM Model 2025*. Brussels: European Foundation for Quality Management.)

sustainable ecosystems. Like Vision Zero, Sustainability is an ambitious long-term goal. Where Vision Zero concentrates on SHW, ecological sustainability focuses on achieving zero climate impact, the development of a circular economy wherein production and consumption generate zero waste and zero emissions, and safeguarding biodiversity. These goals require both technological and social innovations and societal transitions, for instance, our energy systems and patterns of energy consumption. The management of technological development and innovation is key for making significant steps towards ecological sustainability, as is the involvement of key stakeholders.

Sustainability also comprises social aspects, though they often get less attention. The social aspects of sustainability are largely based on social and human values like social justice and equality. Sustainable consumption requires behavioural and cultural changes in society, as well as the elimination of perverse incentives. It is broadly accepted that the fundamental labour rights as defined by the ILO governing council are essential for sustainability. This means that a safe and healthy working environment is part of the quest for sustainability.

Regarding the social aspects of sustainability, SHW experts have developed the concept of 'sustainable work'.[5] It means achieving working conditions that support people in engaging and remaining in work throughout an extended working life.[6] An important challenge is to match the needs and abilities of individuals with the quality of their jobs and to maintain their workability in a changing economic context throughout their career.

In Section 1.4, the UN Sustainable Development Goals 8 (Decent Work and Economic Growth), 3 (Good Health and Wellbeing), 9 (Industry Innovation and Infrastructure) and 12 (Responsible Production and Consumption) were already mentioned as closely related to Vision Zero; Vision Zero contributes to these SDGs and at the same time, the broad acceptance of the SDGs, also in the world of business, contributes to the support from the world of business for Vision Zero.

The management of sustainability is closely related to *Corporate Social Responsibility* (CSR). That implies responsible and ethical business processes. Human and social values are therefore just as essential for CSR as for Vision Zero. The CSR principle of preventing the externalization of undesirable consequences is implying also the prevention of work-related accidents and ill health. In SHW undesirable consequences are often shifted to the workers and their families, as well as to public health care and social security systems. Another important principle in CSR and the management of sustainability is to have dialogues with stakeholders and take the interests and expectations of the key stakeholders into account in important business decisions. Though in CSR the emphasis is often on the external stakeholders it is very clear that workers are crucial internal stakeholders for any organization. A safe and healthy working environment and work processes are therefore not only important for Vision Zero but just as well for CSR.[7]

In business practice, the management of SHW is often closely connected with environmental management. That is most prominent in manufacturing and high-risk industries like (petro-)chemical industries and nuclear power plants. In these industries, any loss of containment of dangerous materials is simultaneously a threat to the environment, public safety and occupational safety and health. Process safety management is thereby often the link between the management of SHW and environmental management. The use of the Proactive Leading Indicators can, slightly adapted help to make environmental management more proactive and innovative, while confirming and strengthening the innovative character of Vision Zero. This is especially relevant in the manufacturing industries and the high-risk industries.

The management of sustainability/CSR and of SHW also have complementary strengths and weaknesses. In CSR it is a generally accepted good practice (and sometimes a legal obligation) to externally report annually the ins and outs of the CSR activities, performances and challenges. That includes several aspects of SHW and employment. The idea behind external reporting is that it is important to be transparent (for generating trust among stakeholders) and that it generates an incentive for the positive intentions expressed in the reports to be put into practice. External reporting for SHW is, however, still limited and should be further developed. The data presented in external reports should be used more consistently by SHW policy makers and researchers.[8]

10.1.3 Synergies with Other Social Goals

The human and social values that are important for Vision Zero (see Section 4.1) are not only relevant for SHW. These values do not lose their meaning beyond the organization's fence when they are genuinely adopted. For instance, from a value perspective occupational safety, public safety, product safety, food safety and patient safety are closely associated though they are relevant in other settings and for other stakeholders. Similarly, occupational health, environmental health and public health are based on similar values. Based on the same or very similar values, synergies can be created with several societal issues.

FIGURE 10.3 The 5-Zero Culture model of the Occupational Risk Prevention Foundation (ORP). (From ORP, (2024) What Is 5Z? https://5zculture.org/en/home/ (Accessed 22 December 2024).)

Like SHW as values, the seven SHW supporting values (Interconnectedness, participation, trust, justice, responsibility, growth and development and resilience) are relevant beyond the workplace. Depending on the setting and the key stakeholders involved they can be understood or operationalised differently. Nevertheless, the SHW supporting values are important for society at large. This implies opportunities for synergies between Vision Zero and several societal initiatives.

Equal opportunities at work and diversity are linked with the fundamental right of no discrimination (a wellbeing issue); they are now key issues in modern personnel management. Equal opportunities and diversity contribute to social justice at work and in society and can strengthen social security. For organizations, they will contribute to Corporate Social Responsibility as well.

The International Occupational Risk Prevention Foundation (ORP), an international independent non-profit foundation dedicated to improvement in the quality of work life and innovation of organizations and entities, has developed a 5-Zero culture approach that comprises zero accidents and zero diseases (similar to SHW) as well as zero waste, zero inequality and zero unawareness. See Figure 10.3.

Zero accidents and Zero diseases are similar to Vision Zero. Zero Waste stands for ensuring sustainable use of ecological resources. Zero Inequality for strengthening teams and organizations through diversity. Zero Unawareness implies transparency, sharing relevant and reliable information, and the development of human talents.[9] These social values and societal ambitions are closely associated with the concept of sustainable work and are fully in line with Vision Zero.

10.2 NATIONAL AND INTERNATIONAL POLICIES FOR VISION ZERO

National and international organizations, including governmental and non-governmental organizations, social security agencies, sector organizations, employers' organizations and trade unions can enable and promote Vision Zero. Apart from its social value promoting Vision Zero at the national level makes sense as the average costs associated with poor SHW are 5.4% of the global GDP. In most countries, the losses range between 3 and 7% of GDP – that implies significant costs and opportunities to save costs.[10] Many costs of poor SHW are shifted to society at large. Alternatives for the standard GDP are in development and focus more on general wellbeing and sustainability.[11] SHW should have a significant position in such broader economic approaches. But at the national and international level as well, SHW is not only a matter of making money, it has an ethical or value dimension as well. This was nicely expressed by Mr Sergio Mattarella, the president of Italy, after a disaster with several fatalities: '*Safety at work is not only a basic legal obligation, but first and foremost a matter of human dignity. ... Guaranteeing safe conditions at work means respecting the life and value of every person*'.[12]

The International Social Security Association (ISSA) has been promoting Vision Zero since 2017. ISSA member organizations are based in more than 160 countries; in more than 140 countries the ISSA and its national partners promote Vision Zero.

The ISSA developed guides with the seven Golden Rules and the fourteen Proactive Leading Indicators (see Chapter 9), as well as a number of related guides. Thousands of organizations from all parts of the world are now Vision Zero organizations and ISSA partners. They use the ISSA Vision Zero materials which are available free of charge in a range of languages.[13] The International Labour Organization (ILO) promotes Vision Zero and concrete SHW improvement programmes in developing countries through its Vision Zero Fund.[14] The *Vision Zero Fund* was created at a meeting of the G7, the first time ever that the world leaders discussed safety, health and wellbeing at work. The European Union has adopted Vision Zero in its strategic framework programme for health and safety for the period 2021–2027 promoting the Vision Zero approach to work-related deaths and making a strong commitment to prevention.[15] The International Occupational Risk Prevention Foundation (ORP) is actively promoting Vision Zero internationally and was already mentioned in the former section.[16]

Several individual countries have adopted a Vision Zero approach. Most well-known are the Vison Zero approaches from Sweden where Zero accident legislation for workplaces has existed since 1998, and Singapore, where the Ministry of Manpower started a Vision Zero movement in 2015. There are also national initiatives that do not explicitly mention Vision Zero but that comprise an integrated approach to SHW combined with the principle of continual improvement of SHW. The most prominent is the Total Worker Health programme of the National Institute of Occupational Safety and Health in the United States of America.[17]

An interesting trend related to Vision Zero is that governments commit themselves to the aim of *zero death at work*, i.e., to prevent all fatal SHW consequences. The European Trade Unions Confederation published in 2021 a *Zero Deaths* at work manifesto, which received signatures of support from Ministers from 11 EU countries, and many members of the European Parliament.[18] Zero death is an important aim of the strategic framework for the Safety and Health of the European Union. It is also adopted by the Netherlands in their OSH Vision 2040.[19] In practice, zero death focuses usually on the prevention of fatal accidents and exposures to carcinogenic chemicals and materials. So far there is little or no attention to the prevention of structural long working hours (+ 55 hours per week, according to WHO and ILO the main cause of work-related death worldwide) or work-related suicide.

> The aim to prevent all work-related deaths is similar to Vision Zero in the area of road safety. For road safety, the aim of Vision Zero is to prevent fatalities and serious injuries. The underlying principle is to save lives and to increase the number of quality-of-life-adjusted years.
>
> Serious refers to irreversible injury, disease or death of one or more workers, and/or significant loss of quality-of-life adjusted years. The focus on the prevention of serious impacts does not exclude the prevention of minor issues, as it remains desirable to prevent all accidents and work-related ill health. But when choices have to be made, the prevention of serious impacts gets priority.

Important topics for proactive national and international policies are the availability of (validated and) practical methods and tools for proactive risk management, the availability of sufficient expertise in relevant sectors of industry and in governmental organizations, and the availability of capabilities for research and development.[20] Proactivity and anticipation require foresight and future orientation. For governmental organizations that implies a responsibility to horizon scanning and trying to mitigate the development of new emerging risks. Emerging risks are usually a consequence of technological, economic, ecological or social developments. These developments can be influenced by national or international policies in order to prevent them to generate major SHW risks. National policies can also ensure funding for research that focuses on the development of preventive strategies and interventions.

The concept of prevention culture was developed at the international level before it was used on the organizational level. In the Seoul Declaration (2008) a national preventive safety and health culture is defined as

> one in which the right to a safe and healthy working environment is respected at all levels, where governments, employers and workers actively participate in securing a safe and healthy working environment through a system of defined rights, responsibilities and duties, and where the principle of prevention is accorded the highest priority.[21]

It is clearly in the interests of policy makers and inspectors to promote and recognize proactive organizational prevention. Investments in prevention create benefits to others than just the investor; that justifies support from governmental organizations. Moreover, governments can promote mutual organizational learning by actively supporting SHW networking that aims at the exchange of experiences and lessons learned in SHW.

Governments can also lead by example by adopting Vision Zero for governmental organizations, by ensuring active governmental SHW procurement, defining proactive policies when new technologies or new organizational arrangements are introduced in governmental or semi-governmental organizations, and practicing the 12 principles of Worker Friendly Design in governmental and semi-governmental organizations.

In many jurisdictions, significant perverse economic incentives exist. This implies that organizations can shift many costs to society at large. The perverse incentives do not stem from OSH legislation as such but from related socio-economic legislation. When market regulation is regarded as the alternative or complement of legislation, it makes sense to eliminate the perverse incentives and make sure that organizations bear the true costs of poor SHW. That might help to transform long-term societal benefits into short-term organizational benefits. Governments can also support SHW innovation in organizations, for instance, by introducing favourable tax arrangements. But incentives have their limitations. Moreover, it is known that a calculative culture is not supportive of Vision Zero.

10.2.1 National and International Regulation

Vision Zero is primarily based on internal motivation, on 'We *want*' good SHW not on 'We *have to*' (compliance). The aims of Vision Zero go beyond compliance. Nevertheless, OSH regulation has three functions in enabling Vision Zero:

First of all, OSH legislation creates a *level playing field*, a minimum level for all organizations and people involved. Good legislation sets the national *rules of the game* for everybody. The law aims to eliminate false competition by organizations that neglect SHW. That requires adequate governmental supervision and enforcement as a complement to policies and legislation.

Secondly, most organizations consider legal compliance as important and monitor developments in relevant legal requirements. Legal SHW obligations therefore have a high communicative value and certainly trigger external motivation for those who are not internally motivated for good SHW. Legislation contributes significantly to the attention of organizations for SHW issues. In organizations, legislation may trigger SHW policies, debates and dialogues on existing and potential SHW problems as well as improvement of available SHW competences.

Thirdly, legislative requirements reflect what is regarded as (un)acceptable in society. Apart from the *letter of the law*, there is the *spirit of the law*. The social values in society are materialized in the legislation and that forms an important ethical norm. That spirit implies that employers and organizations should take responsibility for a safe and healthy working environment and for adequate SHW prevention.

Unfortunately, legislation often reflects old mental models that are not conducive to Vision Zero. For example, the obligation to carry out a risk assessment in existing workplaces, and not in the stages of design and development, seems to reflect the practices of the first labour inspectors in the nineteenth century that went into workplaces and pointed out the most dangerous situations (see Section 4.6.1).

There is also the dilemma of tailoring requirements to the specific circumstances of organizations. That is often useful to reach SMEs, as there is generally insufficient expertise in SMEs to translate more abstract requirements into practical solutions. However, that easily develops into a kind of 'micro-management' wherein too many details are prescribed by law (and potentially enforced). That hinders innovation and often leads to a *tick the box* attitude, thereby destroying potential intrinsic motivation.

In OSH legislation there seems a tendency to impose personal accountabilities on organizational leaders, like the manslaughter law in the UK, in order to coerce compliance. Only in case of conscious and deliberate violation of OSH legislation, such drastic sanctions make sense and can contribute to SHW improvement. Many cases of legal non-compliance are unconscious and unintended and should be regarded as examples of *organizational error*. It is questionable if it makes sense to criminalize the latter. A comparison with human error in organizations can be helpful in this respect. In safety management, there is a lot of attention to human error as a potential cause of accidents. However, most cases of human error occur unconsciously or unintendedly. It therefore makes little sense to criminalise them,[22] and it is not productive for organizations to sanction their employees. In fact, sanctioning and criminalization may have a counterproductive impact; it can easily be experienced as injustice, increase distrust between management and workers, and discourage the notification of errors. It makes much more sense to stimulate learning from human errors. Probably also national legislation that promotes organizational learning makes more sense than further criminalization.

10.2.2 Labour Inspectorates and the Promotion of Vision Zero

The core activity of labour inspectorates is often defined as supervision and enforcement of labour or OSH laws. It is important for labour inspections not to focus exclusively on fines, guilt, liabilities and punishment. The core activity of regulators can also be defined more broadly as guaranteeing minimum acceptable levels of SHW and promoting good SHW. For the self-definition of labour inspectors and their activities this has considerable consequences: are they a kind of policemen, or are they primarily *agents of change* stimulating improvements and innovation? These different roles are not easily compatible in one professional, but in teams of inspectors that should be no problem. The latter interpretation implies that their interventions can also include the promotion of SHW leadership, proactivity, solutions, learning, and attention for SHW in the early stages of design and development or promoting Vision Zero. That is also why ISSA and the International Association of Labour Inspections (IALI) have jointly developed a guide on the promotion of Vision Zero for labour inspections.[23] IALI advocates modern labour inspection to play a dual role in enforcement and engagement. They regard encouragement and engagement as possibly more important than enforcement. The aim is then to support organizations to take greater ownership of processes to sustainably improve the SHW of workers.

The seven Golden Rules and the fourteen Proactive Leading Indicators can be promoted by labour inspections.[24] The Dutch Labour inspection already asks organizations to investigate serious accidents themselves (if necessary, with the help of professional SHW service providers), and to formulate their lessons learned. Guidelines for accident investigation are thereby given. It makes it much more likely that organizational learning takes place. The outcomes of the investigations are, of course, shared with the inspection. This allows the inspection to evaluate the quality of the investigation and to supervise the implementation of the recommendations and to monitor whether structural improvements are realized within a reasonable timeframe. Evaluation showed it clearly stimulates organization learning in the area of SHW. It goes without saying that the labour inspection can always choose to return to the traditional focus on fines, guilt, liabilities and punishment when an organization demonstrates not to be seriously interested in improving SHW.

Authorities can use their status to promote changes in attitudes, policies and practices to the benefit of SHW. They can disseminate good practices, and give information. They can introduce *naming and faming* of organizations that have developed good practices and recognize and award them.

Many labour inspections do not only focus on SHW as such but also on related issues like illegal labour, extremely long working hours and human exploitation. Tax avoidance and precarious labour often indicate a too strong focus on making money only and insufficient responsibility for SHW issues. Human exploitation often goes together with poor SHW.

In their role as change agents, labour inspectors can contribute to network meetings of organizations to explain the importance of SHW, legal OSH requirements and Vision Zero. They can help to disseminate good practices, including the seven Golden Rules and the fourteen Proactive Leading Indicators. They can initiate challenging conversations with organizational leaders to trigger the development of SHW leadership, thereby communicating the business benefits associated with good SHW. They can communicate and disseminate the underlying values important for SHW (the spirit of the law), in order to contribute to the commitment of organizations and their leaders. Furthermore, they can stimulate the generation of SHW synergies in organizations including addressing internal SHW silos.

Labour inspections can promote the development of ownership and a prevention culture through dialogues with organizational leaders and worker representatives. They are in the position to start meaningful conversations about SHW problems, challenges and directions of solutions. Inspections can also communicate their findings to external stakeholders, thereby, of course, respecting confidentiality. Finally, labour inspections could make SHW data gathered during their inspections available for policy makers and research purposes while guaranteeing confidentiality to the organizations involved.

10.3 VISION ZERO IN THE VALUE CHAIN

The social responsibility of organizations for SHW is not limited to their own personnel. To some extent organizations have a responsibility for SHW of their business partners as well. In the case of contractor chains who work on the site of the client, poor SHW practices by the contracting organization can easily affect SHW of the main organization. Supervising or managing the contractor's activities is therefore often integrated into the SHW management of own personnel. But there is an extended responsibility for the quality of work of other business partners as well.

This implies responsibilities for SHW of the personnel of business partners in the respective value chains. The attention for SHW in value chains is on the one hand triggered by disasters like the Rana Plaza collapse (2013, Bangla Dash)[25] and on the other hand by serious safety, health and environmental problems, often found at the ends of the value chain; Involving primary production in mining or agriculture, and demolition, waste treatment and recycling at the end of the life cycle. The ILO reported for example: *Shipbreaking has grown into a major occupational and environmental health problem in the world. It is amongst the most dangerous of occupations, with unacceptably high levels of fatalities, injuries and work-related diseases.*[26] The gradual development of a more circular economy may imply that in the long-range problems associated with demolition, waste treatment and recycling will increase. Vision Zero in value chains is also triggered by the gradually growing importance of Corporate Social Responsibility and recent legal discussions about extended responsibilities for due diligence in supply chains.[27]

10.3.1 Safety Health and Wellbeing Plus (SHW+)

In the process of developing a Vision Zero guide for SHW in supply chains, the concept of SHW+ was developed. SHW+ broadens the scope from SHW to also include the fundamental labour rights and work-related human rights.[28] See Table 10.1 for the

TABLE 10.1
Definition of the concept of SHW+

SHW+ includes SHW and additionally:

- The abolition of child labour. Child labour is almost always detrimental to the SHW of the working child;
- The elimination of forced or compulsory labour. Forced labour is a serious violation of a fundamental human right and is by definition associated with poor SHW;
- The elimination of discrimination in respect of employment and occupation. Almost by definition, discrimination affects wellbeing of the workers who are discriminated;
- Freedom of association and the effective recognition of the right to collective bargaining. For workers it is an essential right to organize themselves in order to collectively protect their SHW;
- The avoidance of extremely long working hours (> 55 hours per week for longer periods). It is the factor most responsible for work-related death[a]
- The entitlement of an adequate living wage. Adequate living wages enable decent living standards and avoid human exploitation. It is a basic condition for the aims of SDG 8: Decent Work and Economic Growth.

Source: ISSA, (2023) Improving Safety Health and Well-being in Workplaces along Global Supply Chains by Vision Zero. Geneva: International Social Security Association.

[a] WHO/ILO, (2021) *Joint Estimates of the Work-Related Global Burden of Disease and Injury 2006–2016 – Global Monitoring Report.* Geneva: World Health Organization and International Labour Organization. Available at: www.who.int/publications/i/item/9789240034945

definition of SHW+. Problems with one or several of the fundamental rights at work are often the main cause of poor SHW.

Signals that indicate that there might be problems with one of these fundamental rights at work, should be understood as a warning signal for associated SHW problems. It is extremely unlikely that an organization can make significant progress in SHW when these fundamental rights are not respected. These interactions inherent to the concept of SHW+ are shown in Figure 10.4.

The concept of SHW+ is becoming increasingly important in a world with growing inequalities, in developed as well as developing countries. It broadens the scope of SHW with aspects of social justice, industrial relations and sustainability. It also underlines the relationship of SHW with inequalities in the labour market and the role of power in society.

SHW+ is an essential concept when dealing with SHW in value chains and for promoting fair trade and social justice. It is crucial for the social responsibility of multinationals regarding their value chains. An increasing number of countries acknowledge that responsibility is an issue for new or potential legislation, also to prevent false international competition; the governments of 37 countries signed the OECD Declaration on Promoting and Enabling Responsible Business Conduct in the Global Economy.[29]

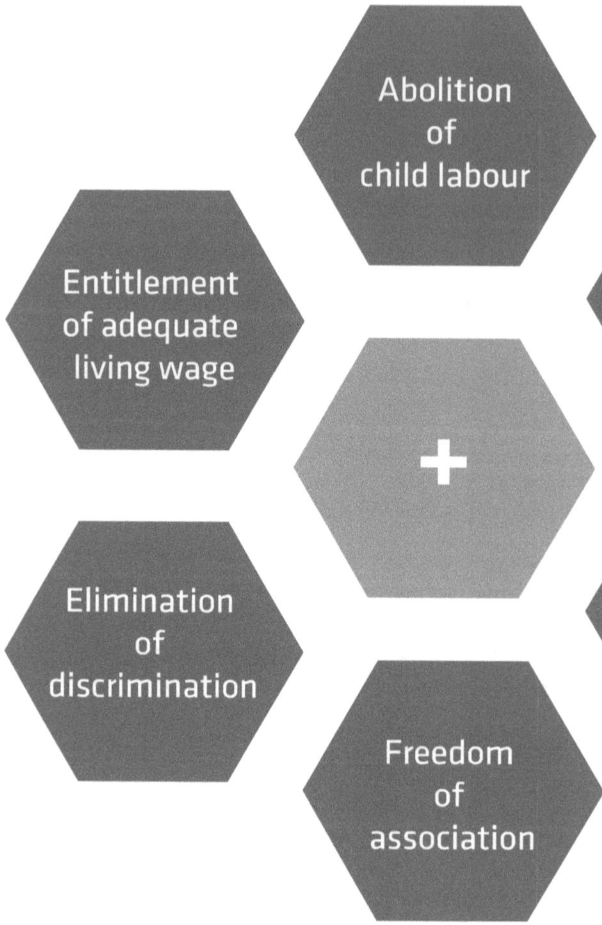

FIGURE 10.4 The Plus of the SHW+ concept.

10.3.2 Developing SHW in the Value Chain

The development of SHW and SHW+ in the value chain is a long-term and complex process. First, it is important to clarify the values, principles and minimum SHW performances that the leading organization regards as essential SHW+ requirements of its business partners. The challenge then is to identify the most relevant value chain partners, i.e. those with the greatest SHW and SHW+ challenges; usually – but not always – to be found at the ends of the value chain. Then it is possible to make contact, to communicate with those partners and to gather information from a variety of sources about the SHW+ challenges. Of course, the organization at the end of the value chain can be asked for information, but usually it is wise to use other information sources too, e.g. from local newspapers, local governments, trade unions, competitors or certification bodies.

The further away business partners are, geographically, but also in terms of the number of chains between the partner and the main organization, the greater the challenge. For contractors and direct suppliers, there are direct contacts and contracts, but organizations have a decreasing *sphere of influence* relative to far-away partners.[30] Another difficulty concerns intercultural differences. What is regarded as normal or acceptable is not the same in the various regions and cultures of the world.

When the contacts with the business partners are good, dialogues around a code of conduct can be organized, knowledge on SHW can be shared, key people can be invited for an internship, etc. It is most important that a process of joint organizational learning and improvement gradually develops.[31] This requires that the partners in the value chain are seen as equals.

10.4 SOCIETAL IMPACTS OF VISION ZERO

Vision Zero focuses primarily on the role of organizations. But it can have a profound impact on people, individually and collectively which is relevant for society at large. Clearly, having work that is good for SHW is a better experience than having work which is affecting SHW or which is only performed for the money. Workers and managers participate in the Vision Zero processes, in the challenge to prevent accidents and work-related illnesses and to continually promote and improve SHW. They have an active role, develop social skills and have valuable experiences associated with personal growth and development.

When people develop ownership of the Vision Zero processes they are involved in, it strengthens their commitment to SHW and the organizational goals. It makes people proud of their work their professionality and on their contributions to achieving Vision Zero. They experience being part of a community with a common goal, which strengthens the bonding to their work, the organization and society at large. Growing trust in the organization and the co-workers makes people emotionally more resilient. They participate and contribute to an organizational culture wherein long-term impacts are taken into account and wherein the wellbeing of people is important. They will realize to be a valuable member of that community and to do meaningful work, which is good for their self-confidence and happiness. That makes people often willing to actively contribute to their community or society at large, including voluntary work.

Meaningful work is an often neglected factor in the support for democracy and the transition towards sustainability. Meaningful work is associated with making a social impact, work that provides both intrinsic and external rewards, helping others, work that entails social interactions, and enjoyment of some aspects of work tasks (such as novelty).[32]

When people do meaningful work, it gives them a deeper fulfilment which is important for a meaningful life. Their life is not characterized by frustration or dissatisfaction; they are workers and citizens with a good work and life. It is not necessary to compensate for the discomforts of work and life with the pursuit of endless material consumption, chasing for image and status or delving into complot theories. Regarding the quest for sustainable consumption, people are only willing to give up something (material welfare, status, short-term benefits) when they have considerable security and self-confidence and experience their (work and) life as meaningful.

Though there is a steadily growing awareness of the need for transition towards sustainability and sustainable consumption, there is also significant social resistance. Transitions are limited by a lack of societal support, resistance to change, and fake news. Contributing to Vision Zero requires reflection on your work and its context, and how you can make a small but significant difference. Is a meaningful working life a blind spot in our thinking about transition towards sustainability?

10.5 CONCLUSIONS

Vision Zero is relevant beyond SHW and organizations for *generating synergies*, for national and international policies, and for *value chain initiatives*.

In practice, SHW is often related to managing Quality. The Zero Defects movement was already mentioned. Synergies between SHW and Quality are most prominent in the concepts of Operational Excellence and World Class Manufacturing. Vision Zero is also relevant for the social aspects of sustainability, and synergies with environmental management. The use of the Proactive Leading Indicators can help to make quality and environmental management more proactive and innovative, while confirming and strengthening the innovative character of Vision Zero.

National and international organizations can promote or enable Vision Zero. That makes socio-economic sense given the *economic losses* associated with poor SHW. Several individual countries have adopted Vision Zero; most well-known are Sweden and Singapore. The development of a national prevention culture is supported by a range of countries. An interesting trend is national commitment to *zero death* at work, similar to Vision Zero in road safety. That puts the focus strongly on the prevention of serious accidents and diseases.

The International Social Security Association (ISSA) and its national partners have been promoting Vision Zero since 2017 in more than 140 countries. The International Labour Organization (ILO) promotes Vision Zero through its Vision Zero Fund. The ORP Foundation promotes Vision Zero internationally as well. Labour inspectors have a role in enforcing SHW legislation but can also act as agents of change. According to the International Association of Labour Inspectors, inspectors should promote Vision Zero in the latter role.

The social responsibility of organizations for SHW is not limited to their own personnel. They also have a responsibility for SHW of their business partners, which is especially relevant at the ends of the value chains. With that purpose in mind, the concept of SHW+ was developed. SHW+ broadens the scope from SHW to also include the fundamental labour rights and work-related human rights.

Vision Zero focuses primarily on the role of organizations. But it can have a profound impact on the people involved, individually and collectively. This may contribute to greater societal support for democracy and the transition to a more sustainable society.

NOTES

1 Moss Kanter, R., (1989) *When Giants Learn to Dance*. London: Simon and Schuster.
2 See for example: Schoër, A., Sochert, R., (2000) *Health Promotion Circles*. Essen: BKK.
3 D'Orazio, L., Messina, R., Schirald, M.M., (2020) Industry 4.0 and world class manufacturing integration: 100 technologies for a WCM-I4.0 matrix. *Applied Sciences*, 10, 4942. Doi:10.3390/app10144942.
4 IBM, (2024) *What Is Operational Excellence?* www.ibm.com/think/topics/operational-excellence. (Accessed 19 June 2024).
5 Kira, M., van Eijnatten, F.M., Balkin, D.B., (2010) Crafting sustainable work: Development of personal resources, *Journal of Organizational Change Management*, 23 (5), 616–632. Doi: 10.1108/09534811011071315.
6 EFILWC, (2024) *Sustainable Work*. Dublin: European Foundation for the Improvement of Living and Working Conditions. www.eurofound.europa.eu/en/topic/sustainable-work (Accessed 20 June 2024).
7 EU OSHA, (2004) *Corporate Social Responsibility and Safety and Health at Work*. Bilbao Research Report 210 from the European Agency for Safety and Health at Work.
8 Jain, A., Zwetsloot, G., Torres, L., (2024) Sustainability, business responsibility and occupational health, safety and wellbeing and the future of work. Editorial. *Safety Science*, 106463. Doi: 10.1016/j.ssci.2024.106463.
9 Jain, A., Zwetsloot, G., Torres, L., (2024) Sustainability, business responsibility and occupational health, safety and wellbeing and the future of work. Editorial. *Safety Science*, 106463. Doi: 10.1016/j.ssci.2024.106463.
10 Takala, J., Hämäläinen, P., Saun, R., Nygard, R.H., Gagliardi, D., Neupane, S., (2024) Global-, regional- and country-level estimates of the work-related burden of diseases and accidents in 2019. *Scandinavian Journal of Work Environment and Health* 50 (2), 73–78. Doi: 10.5271/sjweh.4132.
11 Fioramonti, L., Coscieme, L., Mortensen, L.F., (2019) From gross domestic product to wellbeing: How alternative indicators can help connect the new economy with the Sustainable Development Goals. *The Anthropocene Review*, 6 (3), 1–16. Doi: 10.1177/2053019619869947.
12 RAINews.it (2024) *Morti sul lavoro la strage silenziosa* (death at work- the silent massacre). Available at www.rainews.it/articoli/2024/10/giornata-per-le-vittime-sul-lavoro-morti-strage-silenziosa-che-non-puo-piu-tacere-2912ab36-1988-450c-a441-66a5abda1ca3.html (Accessed 10 December 2024).
13 All ISSA Vision Zero guides can be downloaded free of charge from https://visionzero.global
14 VZF, (2024) Every worker, everywhere, deserves as safe and healthy workplace. Geneva: Vision Zero Fund, International Labour Organization. https://vzf.ilo.org/ (Accessed 22 December 2024).
15 EC, (2021) *EU Strategic Framework on Health and Safety at Work 2021–2027 Occupational Safety and Health in a Changing World of Work*. Brussels: European Commission, COM, 323 final.
16 ORP, (2024) Who We Are, ORP Foundation. https://fiorp.org/en/the-foundation/who-we-are/ (Accessed 22 December 2024).
17 NIOSH, (2024) NIOSH Total Worker Health® Program www.cdc.gov/niosh/twh/programs/index.html (Accessed 22 December 2024).
18 ETUC, (2021) *Manifesto Zero Death at Work*. Brussels: European Trade Unions Confederation. www.etuc.org/en/publication/manifesto-zero-death-work
19 SZW, (2023) *Arbovisie 2040*. The Hague, NL: Ministry of Social Affairs and Employment.
20 Houtman, I., Douwes, M., de Jong, T., Meeuwsen, J.M., Jongen, M., et al. (2008) *New Forms of Physical and Psychosocial Health Risks at Work – A Specific Study Request from the European Parliament under the Framework Contract Health and Safety at the Workplace*, Hoofddorp: TNO (report R08738/031-12152.01.02).
21 ILO, (2008) *Seoul Declaration on Safety and Health at Work*. Geneva and Seoul: ILO, ISSA and KOSHA.
22 Dekker, S., (2011) The criminalization of human error in aviation and healthcare: A review. *Safety Science*, 49, 121–127. Doi: 10.1016/j.ssci.2010.09.010.

23 ISSA/IALI, (2023) *Vision Zero Guide for Labour Inspection*. Geneva: International Social Security Association and International Association of Labour Inspections.

24 ISSA/IALI, (2023) *Vision Zero Guide for Labour Inspection*. Geneva: International Social Security Association and International Association of Labour Inspections.

25 Sinkovics, N., Hoque, S.F., Sinkovics R.R., (2019) Rana Plaza collapse aftermath: Are CSR compliance and auditing pressures effective? *Accounting, Auditing & Accountability Journal*, 29 (4), 617–649. Doi: 10.1108/AAAJ-07-2015-2141.

26 ILO, (2015), *Ship-breaking: A Hazardous Work*. www.ilo.org/resource/ship-breaking-hazardous-work-0 (Accessed 22 December 2024).

27 Recently in the European Union, the Directive on corporate sustainability and due diligence came into force. EU (2024) Directive (EU) 2024/1760 of the European Parliament and of the Council of 13 June 2024 on corporate sustainability due diligence and amending Directive (EU) 2019/1937 and Regulation (EU) 2023/2859. Available at https://eur-lex.europa.eu/eli/dir/2024/1760/oj (Accessed 23 December 2024).

28 ISSA, (2023) *Improving Safety Health and Well-being in Workplaces along Global Supply Chains by Vision Zero*. Geneva: International Social Security Association.

29 OECD, (2023) *Declaration on Promoting and Enabling Responsible Business Conduct in the Global Economy*. Organization for Economic Co-operation and Development. Available at: https://legalinstruments.oecd.org/en/instruments/OECD-LEGAL-0489.

30 Frankental, P., House, F., (2000) *Human Rights – Is It Any of Your Business?* London: Amnesty International and the Prince of Wales Business Leaders Forum.

31 ISSA, (2023) *Improving Safety Health and Well-being in Workplaces along Global Supply Chains by Vision Zero*. Geneva: International Social Security Association.

32 Pitt-Catsouphes, M., Mc Namara, T.K., (2019) Quality of employment and wellbeing: Updating our understanding and insights, in: Burke, R.J. and Richardsen, A.M., (eds), *Creating Psychologically Healthy Workplaces*. Cheltenham: Edward Elgar.

11 Evidence for Vision Zero

11.1 THE CHALLENGE OF EVIDENCE

Everybody is interested in effective approaches and solutions. A logical question that can be asked for each approach is: what is the evidence? Many organizations therefore advocate evidence-based solutions. This raises the question what is good evidence?

According to Ulrich Beck, expert knowledge about risk control is not an *objective science* but often a *social construction*. It is based on assumptions that are not value-free. The implicit assumptions of researchers are based on their mental models and the paradigms underlying the theories that are used. As a result, there are different scientific points of view, and risk experts may have conflicting views. Scientists seldom communicate their underlying assumptions or uncertainties explicitly; reflection thereon seems not necessary as long as the researcher belongs to a relevant 'school of thought' and their papers are accepted for publication. Each school has its qualities and limitations.

Research and evidence can be based on a variety of mental models and paradigms. Kuhn said in this respect: 'The competition between paradigms is not the sort of battle that can be resolved by proofs. The various schools of thought accept different quality criteria, and they will not convince the others. Only the test of time, will decide what paradigm will become dominant'.[1] The various – sometimes conflicting – approaches may undermine the authority of the researchers and the credibility of the evidence. It is not uncommon for scientists to give recommendations beyond the area of their expertise. That can further undermine the authority of scientists and indirectly of scientific evidence.

Regarding the values that underlie research and practices, objective evidence is impossible. Objective evidence is only possible when underlying values and assumptions are clarified and used as the criterion for evaluation, within that framework. The quest for evidence focuses on the effectiveness of programmes, approaches, interventions, etc.

Evidence-based usually means that under certain (controlled) conditions an effect is proven. However, not every proven effect is a meaningful effect. It can also be a minor but scientifically significant effect or an effect that is of minor relevance for SHW. Does the evidence imply good practices that have already been shown to work

DOI: 10.1201/9781003621225-11

TABLE 11.1
Traditional scientific evidence compared to practical evidence

Scientific Evidence	Practical Evidence
Based on experimental research, e.g., Randomized controlled trials	Based on practical experimentation or learning from experience, e.g., Case studies
Theory-based	Practice-based
Potentially universal knowledge	Local and context-dependent knowledge
Ideally the basis for practical implementation	Often precedes scientific knowledge
Useful for 'normal problems'	Useful for 'wicked problems'
High credibility for researchers	High credibility for practitioners

Source: Elaborated on Zwetsloot, G.I.J.M., (2014) Evidence of the benefits of a culture of prevention, in: *From Risk to Vision Zero – Proceedings of the International Symposium on Culture of Prevention – Future Strategies*, Helsinki: FIOH. pp. 30–35.

in the practice of organizations? Or only in academic research? In Table 11.1 some basic characteristics of scientific evidence and practical evidence are compared.

Randomized controlled trials (RCTs) are regularly regarded as 'the golden standard' for scientific evidence. Nevertheless, they have several limitations. In practice, there are seldom controlled conditions, comparable to those in RCTs. Control is most problematic in complex social systems or for issues such as prevention culture. Promoting activities based on RCTs leads to an increase in protocols and standards. It is sometimes the illusion of a best practice and may lead to a greater distance between theory and practice. In health care, the limitations of traditional RCTs are increasingly acknowledged, and there is a development towards *real-world evidence*. Two characteristics are important in this respect: a realistic choice of the research population and using multiple sources of data, not only the traditional clinical data from RCTs.

On the other hand, good practices in organizations are seldom objectively evaluated or compared to useful alternatives. Organizations often prefer to communicate their successes but are reluctant to communicate the limitations thereof or the difficulties encountered. When we asked a variety of people in major hazard organizations about the impact of their activities or programmes to improve safety culture, the responses varied greatly in almost every organization.[2] In fact, most people in those organizations did not have the slightest idea, while the safety professionals who were responsible for the programme usually said the impact had been positive, without mentioning clear evidence – clearly an example of confirmation bias. Practical experiences need good evaluations.

Self-assessments by organizations have strengths and weaknesses compared to assessments from independent outside experts. Self-assessments can be very useful in an improvement process. An advantage is then that the people of the organization formulate the conclusions. This implies a greater likelihood of ownership and that the recommended actions will be realized. But in terms of evidence, self-assessments have two important limitations: (1) it is practically impossible to assess the blind spots in (the culture of) your own organization. Important problems or opportunities for improvement can therefore easily be overlooked. (2) The conclusions of the self-assessment may already beforehand take into account the expectations of senior management. This may lead to smoothening the conclusions. A very critical outcome may even have repercussions for those who carried out the self-assessment.

Fortunately, there is a lot of evidence available based on research on a variety of aspects of SHW and SHW management. In the earlier chapters, evidence is given via the references in endnotes. A rather comprehensive overview by Jaap Paauwe of scientific evidence related to wellbeing at work and the positive impact thereof on business processes became recently available.[3]

11.2 LIMITATIONS OF PRESENT SHW RESEARCH

Most SHW research is carried out in large organizations. That is easy to understand. There is more funding available for research in large organizations, as these are able to financially contribute to the funding. Large organizations frequently have their own SHW experts who facilitate the communication between researchers and the organization. Moreover, it is easier to get the cooperation of one or two large organizations than from a range of SMEs. The SHW challenges in large organizations and high-income countries are, however, usually different from those in medium-sized or small organizations. What works in large organizations is usually not very helpful for small organizations. As a result, there is a shortage of knowledge and evidence on how to tackle the greatest SHW risks in our societies.

Evidence implies that cause-and-effect relationships are clarified. The outcomes of experiments are thereby predicted by an existing theory and some additional hypotheses, and in case of non-experimental research they are compared with suggestions from the literature. Evidence in experimental research is therefore usually evidence of a linear relationship of cause and effect. This explains the lack of research on virtuous processes (see Section 9.3). In a self-enhancing process, linear thinking is of little value because a factor can simultaneously be cause and effect. Non-experimental research is usually limited to identifying correlations, not causal relationships.

Lab studies may generate nice results that may have little value for existing practices. Lab research is typically carried out under ideal conditions. In practice, the contexts are often very different. That can be an issue for exoskeletons or breathing protection that in practice does not fit ideally to the skin of workers (leading to leakages and side inflows). There is a growing awareness that the evidence that innovations contribute to significant improvements may be corrupted by significant stakes and creativity in defining and using test standards and protocols.

The context of social experiments is often neglected in research, though it may have a significant impact. Pawson and Tilley[4] demonstrated that context factors influence the effectiveness of interventions. In experimental research, the context factors are ideally kept identical in the experimental group and the control group; in the comparison, the context can then be neglected. Implementation processes are, however, always influenced by a variety of context factors and processes such as values, trust or distrust, culture, credibility of management in the eyes of the workers, the quality of communication and participation. When several of these factors are neglected as irrelevant context factors, the evidence generated may have little practical relevance.

SHW represents human and social values, associated with the fundamental right to a safe and healthy working environment. SHW research and evidence is, however, strongly focused on the rational, control side of SHW. Consequently, the role of values or emotions of those involved (managers and workers, other stakeholders) is hardly taken into account, though it is not uncommon to have a few sentences in a scientific publication that refer to the supposed role of values. Values are regularly mentioned as an important underlying factor of safety culture, but there is little evidence of their precise role. There is research on the relevance of some specific values and their relevance for SHW, e.g., justice.[5] Generally, there is a lack of research and evidence on the values that are important for SHW.[6]

Often the focus in research is on short-term effects. It is then assumed, mostly implicitly, that every single short-term step towards improvement is part of a stepwise process of continual improvement. Unfortunately, that assumption can be wrong. There are many examples that short-term partial improvements hinder or impede more far-reaching long-term improvements at the system level.

In social science, there is the challenge of distinguishing perceptions from reality. Perceptions are real in their consequences, also when responding to questionnaires. There are, of course, techniques to validate questionnaires or the scales used. But such validations are valid only in the cultures where they were validated for, not for other cultures. This is especially relevant for relatively new (emerging) risks and risks that are clearly value-based such as sexual harassment.

Research into innovative (higher-order) learning processes in SHW is complex and scarce. It is beforehand almost impossible to predict the outcomes thereof. What can you expect when creativity and intrinsic motivation are the main drivers for improvement? How to predict the possible outcomes of proactivity or out-of-the-box thinking? In experimental science, this can be studied at best at a high level of abstraction. The better alternative is often using mixed methodologies, whereby qualitative studies complement experimental science or the use of questionnaires. It implies a revaluation of qualitative research.

11.3 KEY CHARACTERISTICS OF VISION ZERO THAT ARE RELEVANT FOR RESEARCH AND EVIDENCE

For Vision Zero, it is important to 'see the big picture'. However, the world of science is increasingly divided into many subdisciplines and specializations, which makes it more difficult for science to 'see the big picture' and to come up with systemic solutions. That is just one of the difficulties in generating evidence that supports Vision Zero.

Another key feature of Vision Zero is the long-term perspective and strategic approach. Research with a long-term perspective is scarce, mostly because it is expensive. Long-term research also implies more confounding factors that influence the findings; that makes it more difficult to come to convincing conclusions. A good example of a long-term research showing the impact of innovation is the case study of the New Zealand Aluminium Smelters by Young relevant to the Zero Accident Vision.[7] The need for more research on safety and strategy is confirmed by Le Coze.[8]

Synergies are another important aspect of Vision Zero: not only synergies between the area's safety, health and wellbeing but also synergies between SHW and business processes. There are certainly scientific publications that show some causal relationships between the three aspects of SHW, for instance, between wellbeing and safety.[9] Nevertheless, systematic research on SHW synergies more broadly is not yet available. Evidence of the relationship between investing in SHW and good business[10] is limited (see Chapter 3) but suggests opportunities for synergies.

11.4 LINES OF RESEARCH CONDUCIVE TO VISION ZERO

The key characteristics of Vision Zero as mentioned above (to see the big picture, a long-term perspective, the combination of ratio and values and the generation of synergies) are, of course, also relevant for future research that aims to support Vision Zero.

There is a lack of research wherein safety, health and wellbeing at work are jointly the focus. Most research is on only one of these aspects. There are, however, positive as well as negative interactions possible, and more research can shed more light on the potential synergies.

A long-term perspective would require more studies on long-term impacts of interventions, more research on processes of continual improvement and more evaluations of short-term improvements in a long-term perspective. It would also require more research efforts into organizational strategies and (strategic) SHW. When factors that are conducive to long-term improvements are identified, they can also be used to evaluate the expected long-term impact of short-term research.

The importance of SHW as human and social values would require evidence on how values can best be incorporated into rational decision-making as well as research on the impact of the supporting values. Values are associated with ethics and would also imply research on ethical dilemmas and how the various SHW stakeholders (can) deal with them. This should include research on how SHW fits into broader CSR approaches and how to make (better) use of tools and approaches developed for CSR or sustainability.[11] There is also a need for research on the adoption of SHW as values and on potential synergies with social goals like public safety, public health and well-being, equality and so on.

More research on the business ratios and the economics of good SHW would be useful as well. Interesting areas for research are then the impact of economic incentives (including perverse incentives) on the motivation of organizations and their leaders. How can long-term (societal) benefits and non-monetary benefits of good SHW and Vision Zero best be assessed and taken into account in business decisions? What can be learned in this respect from social enterprises that rank social benefits above economic benefits? Does *innovation to zero* deliver new types of benefits for businesses while improving SHW? How can the attractiveness of the labour market further be increased through Vision Zero and better SHW (both for potential new employees and for existing employees)? What are successful ways to convince employers to invest more in prevention already in the early design phase?

The concept of a prevention culture also needs further investigation. What does it mean to broaden the culture concept from safety to Vision Zero? How does it relate to organizational or national culture? What are the roles of leadership, participation, and elements of the Vision Zero philosophy like intrinsic motivation, commitment strategy, etc.?

Furthermore, opportunities for synergies within organizations, including SHW synergies and synergies with business processes are relevant. Research on synergies with similar business aspects such as sustainability, quality, public safety, public health, security, cyber security, etc. is also appropriate. It would require breaking through the traditional silos of SHW disciplines, and beyond. When considering the various forms of synergy, the involvement of new types of SHW stakeholders is interesting as well.

Apart from the topics above which are directly related to Vision Zero, there are also a few other areas for research that are relevant for SHW in general but certainly would support the development of Vision Zero as well. These include combined risks, interactions of rights and strategies to reduce the severity of risks (e.g., by design or by reducing vulnerabilities).

The impact and prevention of combinations of SHW risks (instead of individual risks one by one) is more important than ever before. This implies the need to consider the SHW risks in combination also beyond the existing SHW silos, instead of studying them as independent issues studies through separated disciplines.

Research to develop a better understanding of the relationships and interactions between the fundamental labour rights would be useful too. How do they interact and how far are they mutually dependent? What does this imply for Vision Zero from a value chain perspective? What approaches are most effective in promoting Vision Zero in value chains?

Risk has the dimensions of chance and (potential) impact. In many organizations, most attention goes to the reduction of the chance for accidents and diseases (for safety following the Heinrich pyramid). However, it is necessary to focus on research for minimizing the impacts of events as well. There are two main routes in this respect: through better design, using the 12 principles of Worker Friendly Design (see Section 6.5.5). Several of these principles should be further explored and evaluated for SHW, while the interactions between the 12 principles require attention too. The second route is through reducing vulnerabilities. Traditionally there is attention to the vulnerabilities and additional requirements for young workers and pregnant women and so forth. A main issue is adapting the work to the worker (workplace, tools, work times, etc., also in case of disabilities). Reducing vulnerabilities is certainly an issue for health and wellbeing; for safety, it would imply research on the reduction of (credible) worst-case scenarios. Finally, research on the most vulnerable workers in society would be welcome: for instance, on SHW of migrant workers, minorities, those involved in illegal work and more in general on precarious jobs. We also need more research on SHW in the most vulnerable organizations in society, often small enterprises that are struggling to survive economically. The opportunities implied by the rapid developments in Information Technology also deserve a lot of attention, see Section 6.6. Several topics mentioned earlier deserve more research as well; e.g., to explore successful strategies and opportunities for promoting and enabling proactivity, ownership and intrinsic motivation; to support virtuous processes and innovative learning; and on mental models that are underlying SHW policies (not only in organizations but also nationally and internationally).

Researchers depend on the available budget for their research. The above-mentioned recommendations for research are therefore just as relevant for research funding organizations.

Finally, it is interesting to generate evidence on the added value, or absence thereof, of Vision Zero for the SHW performance of organizations. For this reason, some researchers compared the safety performance of organizations that have adopted Vision Zero with similar organizations that have not adopted Vision Zero. It is a difficult type of study, because it compares snapshots in long-term developments. Organizations can adopt Vision Zero at any moment. A well-known example of this type of study is the Sherratt and Dainty study.[12] They analysed accident data from the construction sector in the UK. Later, a complementary analysis revealed that the difference between the two groups was made by one large organization that suffered from four fatalities in the period of investigation.[13] According to the organization's website, they had adopted a zero-accident approach, while accident investigations showed many shortcomings in their management of safety. When one organization significantly influences the outcomes of such a study, the scientific value of the comparison at the group level is minimal. For creating similar groups, such studies need to take into account the history of the Vision Zero commitment. It is not unusual that organizations start with Vision Zero after being involved in one or several fatal or serious accidents (see Section 9.6). Initially, these are poor performers. It will take several years until their performance is really good. It therefore makes more sense to analyse the development of similar organizations through time and compare the developments in their SHW performances. A related problem with this kind of studies is that some organizations like the Vision Zero slogan, but not the consequence of going seriously for prevention. That can be regarded as an implementation failure.[14] It goes without saying that research on the half-hearted adoption of Vision Zero does not shed any light on the added value of the concept.

11.5 CONCLUSIONS

There is a great interest in evidence. Evidence-based solutions are preferred by most organizations. But what is good evidence? What evidence is available? When is evidence objective? When is it a social construction? What research is needed to enlarge the evidence base of Vision Zero?

It is useful to distinguish scientific evidence from practical evidence. These two types of evidence are complementary in their strengths and weaknesses. While the strength of scientific evidence lies in its stringent methodology in a controlled environment, its weakness is often neglecting context factors, and the impact of human behaviour and economic interests in practice. The strength of practical evidence lies in taking into account such factors, but there is often a lack of rigid evaluation.

Several key characteristics of Vision Zero are relevant for research and evidence: the integration of SHW, a system approach, a long-term or strategic perspective, the value and rights perspective next to the ratio, synergies and the dynamics of resilience. Scientific evidence on these aspects is scarce, while practical experiences are often not well documented and evaluated.

There is already a lot of evidence that supports Vision Zero. Most evidence stems from research on SHW improvements that is not specific to Vision Zero. However, the combination of these elements of evidence – as mentioned in the earlier chapters – clearly underlines the importance and potential of Vision Zero. But there is also a clear need for evidence that takes into account the specific characteristics of Vision Zero as mentioned above. Areas that are relevant for SHW in general and also important for Vision Zero are: combinations of SHW risks, the interactions between the five fundamental labour rights and research to reduce the potential severity of SHW risks, including research to reduce vulnerabilities and strengthen resilience. Finally, let's not forget the many opportunities implied by the rapid developments in Information Technology.

NOTES

1 Kuhn, T.S., (2012) The *Structure of Scientific Revolutions* (50th anniversary edition). Chicago and London: The University of Chicago. p. 147.

2 Zwetsloot, G.I.J.M., Van Middelaar, J., Van der Beek, D., (2020) Repeated assessment of process safety culture in major hazard industries in the Rotterdam Region (Netherlands). *Journal of Cleaner Production*, 257. Doi: 10.1016/j.jclepro.2020.120540.

3 Paauwe, J., (2024) *Progressing Performance and Well-being at Work: Travelling the Loop.* Cheltenham: E. Elgar.

4 Pawson, R., Tilley, N., (1997), *Realistic Evaluation.* London: Sage.

5 Daniels, N., Kennedy, B.P., Kawachi, I., (2002) Why justice is good for our health, in: Fernando Lolas, S., Lorenzo Agar, C., (eds), *Interfaces between Bioethics and the Empirical Social Sciences. Regional Program on Bio-ethics OPS/OMS.* Chile: Pan American Health Organization and World Health Organisation. Dekker, S.W., (2017) *Just Culture.* Boca Raton: CRC Press.

6 To the best of my knowledge the only paper to address this is Zwetsloot, G.I.J.M., van Scheppingen, A.R., Bos, E.H., Dijkman, A., Starren, A., (2013) The core values that support health, safety and well-being at work, *Safety and Health at Work*, 4, 187–196. Doi: 10.1016/j.shaw/2013.10.001.

7 Young, S., (2014) From Zero to Hero. A case study of industrial injury reduction: New Zealand Aluminium Smelters Limited, *Safety Science*, 64, 99–108. Doi: 10.1016/j.ssci.2013.11.016.

8 Le Coze, J.C., (2019) Safety as strategy: Mistakes, failures and fiascos in high-risk systems, *Safety Science*, 116, 259274. Doi: 10.1016/j.ssci.2019.02.023.

9 Nahrgang, J.D., Morgeson, F.P., Hofmann, D.A., (2011) Safety at work: A meta-analytic investigation of the link between job demands, job resources, burnout, engagement and safety outcomes. *Journal of Applied Psychology*, 96 (1), 71–94. Doi: 10.1037/a0021484.

10 Two examples: Bräunig, D., Kohstall, I.T., (2012) *Calculating the International Return on Prevention for Organizations: Costs and Benefits of Investments in Occupational Safety and Health.* Geneva: International Social Security Association. Tompa, E., Mofidi, A., van den Heuvel, S., van Bree, T., Michaelsen, F., Jung, Y., Porsch, L., van Emmerik, M., (2021) Economic burden of work

injuries and diseases: A framework and application in five European Union countries, *BMC Public Health*, 21, 49. Doi: 10.1186/s12889-020-10050-7.
11 Jain, A., Zwetsloot, G.I.J.M., Torres, L., (2024) Sustainability, business responsibility and occupational health, safety and wellbeing in the future of work (editorial), *Safety Science*, 174, 106463. Doi: 10.1016/j.ssci.2024.106463.
12 Sherratt, F., Dainty, A.R.J., (2017) UK construction safety: A safety paradox? *Policy and Practice in Health and Safety.* Doi: 10.1080/14773996.2017.1305040.
13 Difford, P., (2020) *Observations on Zero Harm and the Management of Health and Safety in Construction, Part 2.* Available at: https://neucom.uk/zero-harm-safety-management-and-safety-differently/ (Accessed 16 December 2020).
14 Zwetsloot, G.I.J.M., (2017) Vision Zero: Promising perspectives and implementation failures; A commentary on the papers by Sherratt and Dainty and Dekker, *Policy and Practice in Health and Safety*, 15(2), 120–123. Doi: 10.1080/14773996.2017.1373998.

Index

A

Accident investigation, 74, 100, 134, 180, 188, 192, 207
Ageing, 12–13, 42, 49, 122, 162
Ambiguity, 16–18, 21, 38, 57, 65, 97, 113, 124, 128, 130, 158, 162, 187, 194
Anticipation, 20, 70, 76, 80–81, 130, 134, 136, 166, 205
Assessments, *see* Risk assessment; Self-assessment
Assumption, 13, 15, 20, 82, 86–91, 160, 215
Automation, 15, 19, 43, 95
Autonomy, 19–20, 26, 56, 66, 69, 74–75, 80, 90, 101, 103–104, 114, 120, 132–133, 137, 143, 181, 193

B

Beck, U., 13
Beer, M., 63
Behaviour, 14, 18, 20, 30, 32, 39, 51, 56, 58, 60, 63, 65, 69–73, 78, 80–81, 84–85, 90, 95, 97–100, 103–106, 111–114, 118, 120–125, 132, 136–137, 140–150, 154, 162, 166–167, 178, 185–190, 222
  behavioural discipline, 16, 83, 95
  exemplary behaviour, 71, 73, 98, 100, 114, 141, 154, 185–186
Blind spots, 14, 85, 107, 112–114, 140–141, 150–155, 192, 212, 217
Burnout, 65, 80, 110, 158, 177
Business
  business case, 47, 50, 53
  business model, 12–13, 80, 110, 129
  business opportunity, 110
  business partners, 41, 49
  business performance, 42
  business resource, 19, 29, 46–47
  business risk, 48, 108, 111, 169
  business value, 41

C

Change
  agents of change, 207, 212
  cultural change, 13, 57, 73, 95, 141
  demographic change, 12, 30
  ecological change, 13
  economic change, 7, 11–12
  organizational change, 19, 108–109, 119, 188
  technological change, 12, 81
Child labour, 3–5, 8, 51, 209, 220
Commitment
  commitment strategy, 55–58, 66, 68, 81, 99, 161, 174, 177, 182, 220
  leadership commitment, 39, 56, 63, 71, 99, 141, 179, 198
  personal commitment, 57, 96, 98–100, 110, 114, 147, 149, 179
Communication, 12, 15, 17, 19, 29, 38, 56–57, 65–66, 69, 71, 81, 87, 90, 96–97, 99–100, 106, 111–114, 131, 135, 144, 147–149, 151, 154, 162, 188, 193, 217–218
Competence, 17, 28, 56, 74–76, 90, 98, 105, 112–113, 166, 175, 179, 184, 198
Complacency, 38, 65–66, 106, 190
Complexity, 13–17, 56–57, 79, 97, 122, 128–131, 136, 158, 190
Compliance, 1, 6, 23, 38, 42–43, 46, 54–57, 60, 74, 79, 81, 85, 96, 98–101, 120, 123, 131, 137, 142–143, 153–155, 159, 163, 165–169, 171, 174, 178, 183, 185, 192, 205–206
Context, 11, 17, 30, 32, 40, 42, 68–70, 74, 77–78, 83–84, 90, 104, 111, 120, 123, 142, 145, 149–150, 161, 163, 165, 180, 201, 212, 216–218, 222
Continual improvement, 23–24, 33, 37, 40, 52, 54, 63, 66, 134, 148, 157, 159, 161, 164–165, 168, 178, 183–184, 204, 218–219
Cooperation, 12, 17, 19, 20, 44, 54, 56, 71, 87, 90, 149, 162, 187, 217
Corporate social responsibility (CSR), 5–7, 31, 41, 45, 59, 71, 199–203, 208, 219
Creativity, 14, 17, 20, 28–29, 41, 43, 46, 49, 54, 56, 58, 60, 70, 104, 111, 120, 122, 131, 147, 162, 166, 187, 193, 217–218
Credibility, 44, 53, 59, 65, 71, 99, 148, 215–218
Culture, *see* Just culture; Prevention culture; Safety culture; Us *vs.* them
Costs, 19, 36, 40, 45, 76, 85, 124–126, 130, 135, 145, 152, 163, 167, 189, 203
  costs for the employer, 6, 8, 45–50, 83, 107–108, 111, 118–119, 122
  cost for the employee, 85
  costs for society, 7–8, 49, 167, 205

D

Decent work, 6–7, 9, 31, 37, 69, 201, 209
Decision-making, 19, 22, 37, 40, 42, 55, 76, 79, 81, 87–88, 105, 123, 126–131, 141, 143, 165, 178, 219

De Geus, A., 72
Deming W. E., 52, 165
Deming cycle, 87, 164–165; *see also* Continual improvement
Design stage, 27, 50, 76, 105, 114, 125–127, 173, 175
  design thinking, 127–128, 137
  see also Worker friendly design (WFD)
Dialogue, 20, 67–71, 75, 81, 88–90, 96–100, 103–105, 114, 128, 145, 149–150, 155, 166, 176, 179–180, 191, 201, 206, 208, 211
Digitalisation, 12, 15, 24, 135, 188
Dilemma, 52, 182, 206
Direction, 32, 36–37, 56, 58, 63, 98, 113–114, 148, 173, 179, 198
Discrimination, 3–6, 19, 43, 51, 69, 80, 112, 136, 203, 209
Diversity, 12, 17–19, 69, 122, 203
Don'ts, 192–194, 222

E

Early warning, 70, 80, 106, 108, 130, 137, 144, 187–188, 194
Emergence, 17, 19, 21, 28, 32, 40, 46, 90, 113, 131, 174, 190
  Emergency, 82, 97, 121
Employability, 45, 49, 70
Employment, 1–5, 19, 36, 41, 49, 51, 75, 96, 101–104, 141, 201, 209
  employer of choice, 36, 41, 45, 49, 66, 71, 90, 190
  employment relationship, 101, 104
Engagement, 22, 24, 26, 38, 56–57, 64, 69, 71, 73–75, 81, 90, 104–105, 114, 207
Ehnes, H., 171
Ergonomics, 7, 40, 50, 95, 130–131, 136–137, 175, 192
Error
  human error, 18–20, 28, 32, 78, 84, 109, 130–131, 133, 136, 160, 166, 190, 192, 206
  organizational error, 206
Ethics, 5–6, 15, 23–24, 30–32, 37, 44, 52, 56, 71, 99, 102, 105, 108, 201, 203, 206, 219
Evidence, 26, 36, 73, 123, 128, 130, 133, 177–178, 185, 215, 217–222

F

Fatal accident, 3, 25, 39, 44, 74, 94, 101, 110, 178, 191, 194, 204, 221
Fear, 101, 155, 158
  absence of, 57, 64, 69, 75, 95, 145, 155
  see also Psychological safety
Feedback, 30, 57, 69–70, 75, 86, 89, 103, 105–106, 111, 127, 134–135, 137, 146–151, 174, 176–177, 182, 184–186
Flow, 26, 80
Forced labour, 4, 5, 8, 209
Freedom of association, 3–5, 8, 209

G

Goal setting, 50–51, 59, 78, 103, 174
  goal setting theory, 36, 56
Golden rules, 177–182
Good enough, 79, 158–159
Gross domestic product (GDP), 7, 9, 203

H

Hazards
  psychosocial hazards, 22, 85, 119, 129
  technological hazards, 85, 119, 132, 175
Health
  mental health, 1, 7–8, 11–14, 24, 26, 29, 32, 43–46, 49, 65, 99, 110, 120–121, 129, 181
  occupational health, 1, 32, 82–83, 120, 177, 202
  physical health, 121
  positive health, 24, 33
  social health, 29
  see also Total worker health
Hierarchy of prevention measures, 117–124, 136
High-performance, 56, 103
High reliability organizations (HRO), 26, 38, 80, 187–188
Hollnagel, E., 27–28, 89, 123, 190–191
Human factors, *see* Ergonomics
Human resources, 14, 29, 46, 57, 160, 162
Human and social values, 31–33, 37–39, 54–55, 59, 63–64, 90, 147, 201–202, 218–219
Human toll, 2, 58, 117, 222

I

Incentives, 50, 56, 100, 147, 149, 152, 167, 200–201, 205, 219
  perverse incentives, 200, 205
In control, 27, 81, 96–97, 106, 114
Indicators
  lagging indicators, 25, 52, 177–178
  proactive leading indicators, 178–181
Information Technology (IT), 12, 15, 18–20, 29–30, 81, 89, 109, 119, 182, 221
Injury, 2–3, 8–9, 21, 24–26, 29, 31, 39, 48–49, 52, 177, 204, 208, 221
Innovation
  Innovation in the system, 159–160, 168
  innovation of the system, 159–160, 168
  Innovating to zero, 53, 63, 160, 168
  workplace innovation, 38, 160, 181
Intangibles, 38, 45, 48, 76, 111, 141–143, 154, 164
  vs. tangibles, 44, 141–142, 164, 167

Interconnectedness, 64, 66, 69, 71, 75, 90, 203
International Electrotechnical Commission (IEC), 161, 175
International Labour Organization (ILO), 1–5, 13, 31, 37, 51, 64, 200, 204, 208–209, 212
International Social Security Association (ISSA), 39, 171–173, 203–204, 207, 212
International Standardization Organization (ISO), 21, 24, 117, 164–165, 175, 177, 183, 194, 197
Investing
  in people, 172, 176
  in SHW, 42, 48–50, 107, 111, 187, 219

J

Job
  job demands, 74, 121
  job control, 74, 80
Juran, J. M., 52, 76–77, 153
Just culture, 57, 65, 105, 133

K

Kai Zen, 198
Key words, 69–70
Kletz, T. A., 119, 126, 129, 134, 166
Kolb, D., 86, 135, 165
Kuhn, T., 215

L

Leadership, 94–116
  leadership challenges, 105–112
  leadership styles, 97, 123, 147
  limitations of, 112–114
Lean, 53, 108–109, 158, 198
Learning
  adaptive learning, 27, 86–91, 169
  collective leaning, 54–56, 81, 86–91
  experiential learning, 87
  individual learning, 86, 88
  innovative learning, 27, 86–91, 154, 169, 221
  learningto learn, 168
  learning on the job, 17, 20, 66, 70, 75, 86, 108, 143, 176, 181
  organizational learning, 32, 46, 54, 80–81, 86–89, 98, 105, 113–114, 151, 179, 183, 194, 205–207, 211
    unlearning, 88
Legislation, 39, 82–85, 91, 101, 110, 112–113, 128–129, 163, 171–172, 204–206, 209, 212
License to operate, 43, 49, 57, 96, 123
Loss aversion, 25, 52
Lost time incident, 21, 111
Loyalty, 19, 66, 69, 98, 102–103, 122

M

Managementof change (MoC), 109, 145, 181
Mattarella, S., 203
Meaningful work, 22, 63, 70–73, 102–104, 211
Mental model, 42, 48, 72, 82, 86, 127, 152
Migrant worker, 6, 8, 12, 220
Motivation
  external motivation, 60, 74, 85, 101, 153, 155, 166–167, 206
  integrated motivation, 73, 90
  intrinsic motivation, 54, 56, 58, 60, 72–74, 79, 81, 90, 98–99, 105, 122, 143, 147, 153–155, 166–167, 176, 185, 187, 193, 206, 218, 220–221
  identification as motivation, 73, 75, 90, 104

N

Negatives, 21, 24–28, 33

O

Occupational Risk Prevention (ORP), 202–204
Operational excellence, 168–169, 198, 212–213; *see also* World class manufacturing
Optimization, 32, 46, 85–86, 106, 109, 111, 133, 136–137, 141, 145, 157–161, 164–169
Options, 76, 119, 123, 126–128, 131, 134, 136–137, 152, 154–155, 179, 183, 185
Outsourcing, 20, 43, 122, 141
Ownership, 38, 54, 56–60, 66, 69–70, 72–75, 79–80, 90, 122, 126, 131, 133, 147, 149, 153–154, 166, 169, 189, 193, 198, 207–208, 211, 216, 221

P

Participation, 17, 20, 22, 33, 36, 38–39, 54, 57, 60, 64, 66, 68–69, 71, 74–75, 80, 84, 87–88, 90, 102–103, 120–121, 128, 131, 134, 137, 145–146, 149, 160, 162, 172, 174, 176, 182, 198, 203, 218, 220
Pensioning, 7–8, 13–14, 132, 167
Perfection(ism), 79, 158–159
Prescription, 23, 80, 104, 122, 142–143, 153, 159–160, 162–164, 166, 190, 193
Personal commitment, 57, 96, 98–100, 110, 114, 147, 149, 179
Planning, 16–17, 21, 38, 45, 49, 57, 64, 76, 78, 80, 87–88, 98, 106, 120, 124–129, 147, 156, 159, 164, 178–181, 193
Popper, K., 28
Positive psychology, 26, 33
Prevention culture, 32, 37–39, 53, 56, 58–60, 64, 90, 98, 105, 113, 123–125, 133, 140–155,

157, 160, 165, 167, 172, 174, 182, 188, 192, 194, 198, 205, 208, 212, 216, 220
Prevention
formal *vs.* informal prevention, 141–144, 153–155
primary prevention, 117, 120–124, 137
Preventive control, 125, 134–137, 173
Pre-work briefings, 180, 183
Proactivity, 9, 20–21, 25, 32, 36–37, 40, 46–47, 52–60, 66, 70, 73, 76, 79–81, 90, 98, 104, 110, 113–114, 117–137, 140–144, 147–155, 165–166, 169, 171, 174–185, 188, 192–194, 198, 201, 204–207, 212, 218, 221
Productivity, 1, 12–13, 17, 29, 24, 28–29, 46–52, 63, 78, 88, 102–103, 107–110, 114, 123, 127, 129, 145, 152–153, 157–158, 169, 181, 187, 190, 193, 206
Priority, 21, 23, 37–38, 50, 57, 63–64, 72–73, 84, 107–108, 111, 117–123, 161, 164–165, 175, 179–180, 189, 204–205
promotion of SHW, 31, 33, 121–123, 207
health promotion, 89, 166, 197
Psychological contract, 102–104, 109, 114, 141
Purpose, 22, 26, 56–57, 63, 70–72, 77, 101, 130, 135, 137, 160, 187, 192, 198, 212

Q

Qualification, 14, 19, 30, 141, 159, 162, 175–176, 179, 182
Quality, 49, 75, 88, 96, 104, 108, 124–126, 132, 145, 148, 154, 157–159, 181, 197, 220
quality management, 38, 76, 102, 124–125, 197–200
quality of life, 2, 8, 101
quality of work, 26, 42, 45, 49, 83, 101–102, 201, 203, 208, 212

R

Relatedness, 74–75, 90
Reputation, 6, 41, 43, 46, 48–49, 59, 98, 111, 123
Resilience, 20, 24, 28, 33, 49, 64–66, 70, 72, 90, 117, 121, 203, 222
Resource
resource view, 44–47
organizational resource, 71
personal resource, 71, 74
resource theory, 29, 44
see also Business resource
Responsibility, 4, 17, 22, 30, 33, 37, 46–47, 54, 64–66, 83, 87, 90, 94–95, 102, 106, 121, 129, 131, 133, 146, 153, 175–176, 199, 203; *see also* Corporate social responsibility
Rights, 1
human rights, 5, 37, 51, 64, 71, 110, 136, 208–212
labour rights, 3–6, 8, 59
Risk
emerging risk, 22, 28, 38, 122, 125, 134, 137, 145, 205, 218
risk control strategy, 57, 161, 164, 215
Risk assessment, 19–20, 30, 109–110, 125–130, 136, 141, 159, 173–181, 206
dynamic risk assessment, 19–20, 30, 109
first minute risk assessment, 109–110, 126, 130
Robot, 15, 19, 30, 132, 135–136
Roggeveen, V., 188
Rousseau, D., 102

S

Safe and healthy work environment, 1–4, 8, 31, 37, 59, 64, 140, 201, 205–206
Safety
collaborative safety, 30
safety culture, 38, 46, 57, 73, 80, 140, 145, 153, 216, 218
new view of safety, 28, 122, 166, 190
occupational safety, 1, 11, 31, 38, 82, 111, 117, 120, 123, 164, 201–202, 204
psychological safety, 57, 69, 74–75, 145–147
road safety, 38–39, 53, 204, 212
Safety-II, 26–30, 33, 76, 89
Salutogenesis, 26
Scenario, 40, 48, 111
Self-assessment, 151, 171, 216–217
Self-determination theory, 56, 74
Scientific management, 12, 16–17, 83, 85
Senge, P., 67, 82, 86
SHW+, 208–212
Sickness absence, 6, 24–25, 29, 47–49, 177, 185, 189
Silo, 14, 20, 33, 54, 82, 85, 88, 90–91, 154, 193, 208, 220
Singh, S., 53, 160
Social capital, 20, 29, 44, 46, 57, 69–70, 72, 88, 104, 145, 191
Social support, 17, 56, 64–65, 69, 74–75, 80, 88, 103–105, 114, 147, 152, 176, 184
Socio-technical
socio-technical design, 17
socio-technical school, 17, 129, 132
socio-technical system, 14–15, 18–21, 30, 110, 122, 145, 157–163, 168, 171, 174
Spit, P., 94
Stagnation, 7–8
Standards, 21, 23, 57, 82, 84, 91, 95–96, 117, 129, 158, 160–167, 175, 177–178, 183, 197, 209, 216–217
standardization process, 16, 119, 152, 160–164, 173, 190

Suggestions for improvement, 57, 65, 80–81, 109, 145–149, 151, 178, 182
Sustainability, 7, 13, 38, 43, 64, 71–72, 124–125, 154, 157, 199–203, 209, 211–212, 219–220
  Sustainable Development Goals, 6–9, 13, 51, 201, 209
Surprise, 21, 86, 95, 105, 122–123, 155, 158, 188–189, 194
  unplanned events, 26, 49, 106, 179–180
Synergy, 11, 33, 38, 42, 50, 54–55, 60, 63, 66, 91, 153–155, 167–168, 178–179, 192–193, 197–198, 202–203, 208, 212, 219, 222
System
  production system, 101, 131, 135–136, 161, 163, 169, 174–175
  see also Socio-technical system

T

Taboo, 97, 112, 140–141, 150, 152, 155, 192
Team performance, 162, 187
Theory
  theory U, 160
  theory W, 190–191
  theory Z, 190–191
Total Worker Health, 31, 204
Trade union, 1, 5, 31, 85, 113, 203–204, 210
Trust, 17, 20, 49, 57, 64–66, 69, 71, 75, 81, 88, 90, 104, 107, 114, 128, 131, 142–145, 147, 155, 163–164, 185, 188, 193, 201, 203, 211, 218
  distrust, 13, 65, 103–104, 112, 143, 163, 206, 218
Tunnel vision, 65, 150, 193

U

Uncertainty, 13–15, 21–22, 25, 28, 50, 57, 97, 119–122, 128, 130, 169, 184, 215
UNICEF, 4
Us *vs.* them, 103, 112, 189

V

Value(s)
  aligning, 66, 67
  business value, 41–42, 59
  value case, 99
  value chain, 11, 41, 96, 123, 208–212, 220
  core value, 37–38, 44, 50, 63–66, 69–70, 77–78, 110, 146, 155
  human and social values, 31–33, 37–39, 54–55, 59, 63–64, 90, 147, 201–202, 218–219
  shared values, 57, 68–71, 154, 157, 169, 185, 188
Variability, 26–27, 30, 56, 76, 158, 181, 190–191
Vicious processes, 19, 184–185, 191
Virtuous processes, 184–187, 194, 217
Vision Zero Fund, 204, 212

W

Walk arounds, 107, 149, 155
Weak signal, 80, 106, 130, 152, 187–188, 190, 194
Wisdom, 33, 89, 99, 176, 190–194
Work
  division of work, 11, 17, 83, 85, 87–88, 132
  as done, 113, 141, 143, 164–165, 176, 190
  hazardous work, 4, 8
  as imagined, 113, 141, 143, 164–165, 176, 190, 194
  intensification of work, 7, 11, 14, 18, 44, 108
  meaningful work, 22, 63, 70–73, 102–104, 211
  precarious work, 4, 6, 12, 14, 20, 102
  work engagement, 38, 56, 73–75, 81, 90
  work environment, 8, 23–24, 26, 33, 37, 64, 121, 136, 142, 198, 206
  work organization, 14, 17, 19–22, 26, 66, 74–75, 80, 85, 88, 90, 102, 107–108, 119, 122–123, 125, 128–129, 131, 133, 137, 146, 160, 166, 173, 176–177, 180–181, 185, 191, 193
  work place induction, 105
  see also Quality of work
Worker friendly design (WFD), 129–137, 175, 193, 205, 220
Work-related accidents, 3, 25, 39, 43, 52, 58–59, 81–82, 177, 201
  work-related death, 2–3, 53, 129, 204, 209
  work-related illnesses, 3, 8, 30, 43, 73, 177–178, 211
World class manufacturing (WCM), 29, 45, 51, 198, 212
World Health Organization (WHO), 2–3, 29, 204, 209

Z

Zero
  5 zero culture, 202–203
  5 zero manufacturing, 198–199
  zero accidents, 29, 31, 38–39, 45, 51, 73, 94, 96, 100, 198, 203
  zero defects, 38–39, 51–53, 197–198, 212
  zero death, 204, 212
  zero production losses, 6, 8, 173, 198

FSC
www.fsc.org
MIX
Papier | Fördert gute Waldnutzung
FSC® C083411

Kolibri 360 GmbH
Ferdinand-Jühlke-Straße 7
99095 Erfurt, Deutschland
produktsicherheit@kolibri360.de